DELIUS KLASING

BULLI LOVE

DELIUS KLASING VERLAG

THE SAMBA OWNED BY Yanto Widodo parks in a tea plantage on the island of Java surrounded by tea pickers – a picture just like that found six decades ago. Mobil Rakyat – that's what you say in Indonesia when you're talking about a Volkswagen

SOMEWHERE IN NO MAN'S LAND – Andreas Lidl has driven his T3 Syncro to many destinations with natural settings such as Africa or the vast expanse of Asia; here, for example, far from Ulan Bator in Mongolia

LOVE STORY

No other automobile has inspired so much love across multiple generations or given its owners as much freedom as the Bulli – a love story with a perpetually happy ending

Things have no soul, says the thinking man, but then hesitates immediately as he senses himself that there is in fact more than can be explained by reason alone. Naturally, a Bulli – the Volkswagen Type 2, be it a T1 or a T6 – is in principle just a thing – yet at the same time we know that this is mere theory. For it is much more, it is a part of us. A godsend of automotive evolution. While normally cost and utility alone are the deciding factors for continuation of a species in the Darwinism of the street, the Bulli rises above a mere logic of numbers – and not just because of its sheer size.

We are the Bulli. It is our transporter into new worlds, emotionally and geographically. It is our projection of a life of possibilities. It is companionship and freedom. It opens its doors and sets off for the destinations of our dreams, whatever they may look like.

Each of us associates something with the Bulli; it is part of our collective memory. As children, we sat in the back and travelled to exciting new shores. After thousands of kilometres, or just a few, its door opened to broaden the view to horizons that got engraved on our memories. Whether it was the football pitch of the neighbouring village or a mountain lake in the Alps or Sweden's never-ending woods.

But with the Bulli that eternal place of yearning was never just the destination; the journey itself was already a pleasure in it. We got up to nonsense, played games, dreamt, read and above all talked and listened. As children we often slept in its shelter and arrived safe and sound. As wild youth we tasted freedom, without giving a thought to where we were headed. It had everything – a place to sleep and all that you needed to see the world and explore its every last corner.

On the road in a VW Bulli from childhood

Of course, surfing years were Bulli years, too. Of course, half the kindergarten swarmed forward to ride in the Bulli. Of course, it was on hand without complaint for every flat move, especially the first few. Always a friend that lends a hand, a mate that shares adventures with you, a great love that you never forget – or who perhaps is even still around. With a perpetually happy ending.

TEXT: MAX B. OERTEL

Naturally, the VW Type 2 is not just about emotion and image. It is useful, practical, has staying power, holds its value over many years, and offers a lot – and yet it is almost impossible to describe what it means to ride in it, to travel with it. You sit above it all. You are away from the hustle and bustle, elevated. A sovereign – without equals. All of that is, of course, not rational – fortunately. As being sensible often feels too much like the opposite of sensual.

A member of the working class – yet classless

The Bulli was the first true vehicle of its type in automotive history. In the early days, it was almost a political statement. A kind of motoring adversary to everyday work's lack of freedom – despite itself being a born worker! It never placed any importance on status, and yet with every kilometre, every year and every generation, its reputation, its status as an icon has grown and grown.

That's not something you can plan. Countless strategists at other brands have tried in vain. Now, in the sixth change of nomenclature, the competition is – with respect and due distance – following in its tracks. Bragging now and then about supposedly being able to do everything better than the original. We hear the words, but we can't feel the love.

What is required to achieve the status of an icon is probably calm composure above all. And experiences that write life stories, and places that leave us with smells and images – and the laughter of children, friends or our sweetheart beside us.

It is these eternal moments that continually revitalise this unique Volkswagen and give it its aura, without any attitude at all, just as it is. Quite naturally, they are the sounds of this love story – to be continued. ❤

THE EIGHT-LANE CABLE-STAYED BRIDGE IN MUMBAI has been given the name Rajiv Gandhi sealink – Santaji Shirke enjoys the mix of the monotonous vibration of the automotive drumbeat behind him and the breaking of the waves in the bay of Mahim beside him. Small moments of happiness on the couch of his T2 Camper

WESTFALIA
Hella
Hella
MH02CP5464

CONTENTS

THE FROG, CLARA, AND THE ROOF IN THE RIVER

Bevan Beattie of New Zealand once toured Europe in a frog-green Bulli – and he fell in love with the bus. Back home, Bevan and his company Classic Campers now offer other adventure seekers an opportunity to follow their dreams with a whole range of Bulli buses that they too can fall head-over-heels for

On the road along the wild west coast of New Zealand's South Island

KAITERITERI
6 km
GZ 3448
HO 7648

Free camping in Kaikoura. Perfect lighting. As whales swim by, lobsters and dreams are captured

A story should never be told in chronological order. That would be boring. The following story is different, though. If you don't believe it, then read on. It all started in the summer of 1986. Bevan Beattie was 16. He lived on Hawkes Bay on the eastern part of the North Island in New Zealand, and he urgently needed a car. One day, while getting off the school bus, he saw a Beetle with a "For sale" sign. "I borrowed 1,025 dollars from my grandparents and brought my VW home", he recalls. "I then mowed lawns to pay off the debt. My father and I both loved to tinker with cars, so we jazzed up the Beetle." Later, Bevan used the car to chauffeur kids to youth centres – "but the Beetle was simply too small. That's when I got the idea of buying a VW bus.

I looked through the ads in the papers. One day, there was just one newspaper left in the shop in our village, and a woman and I both wanted to buy it." Bevan offered to let her have the paper if she would let him keep the classified section. He explained to her that he was looking to buy a Bulli. "And then the woman told me her husband was trying to sell one", Bevan says with a smile. "It was a nine-seater from 1971 and she said I could have it with or without the engine. I didn't have much money, so I decided to buy the bus without an engine and then fitted it with the 1.6-litre engine from my Beetle." Bevan drove the bus for many years, frequently replacing the engine to get more power. One day, he actually beat a friend in a Subaru Legacy 4WD in a drag race on a cow pasture. "That Bulli is now in my barn waiting for a three-litre Boxer engine", Bevan says with a grin.

On the road with a screwdriver as a shift lever

Bevan completed a degree in auditing in the early 1990s. Like many other young "Kiwis", he then set out for his big "overseas experience" to get to know the world beyond the "down under". In keeping with another tradition, he and a friend bought a Bulli camper in Europe, shelling out 700 pounds for the bus. "The bus was in bad shape and painted in a nauseating frog-green colour", Bevan explains. No matter – they embarked upon their tour of Europe anyway.

"Things came to a screeching halt at the border of Greece", Bevan recalls. "We didn't have the proper papers, and the border guards thought our VW might be stolen. Unbelievable! I mean, who would steal a VW bus in that horrible colour? In the end, they allowed us in for two weeks." Throughout all the months and kilometres covered during the trip, the "frog" only caused trouble once, in Dover: "The gear-stick broke, so I had to shift with a screwdriver, but I could only shift into second or fourth. We drove like that all the way from the coast to London."

During these and other escapades in Europe, Bevan came up with an idea: "I loved going camping, and cars, but most of all Bulli buses. So, I thought to myself that it would be great to offer VW campers for rent for trips in New Zealand." The country's vast landscape and breathtaking scenery, its extensive network of camping sites and the Bulli as a box full of youthful dreams of freedom – that was to be Bevan's business. The name: Classic Campers. Fast-forward to 2003. One day, while on his way to work in Auckland, Bevan stumbled upon the first Bulli that would become part of his fleet later on. "There was a phone

Golden days in Golden Bay: a yellow 1974 Bulli and the Westfalia whose roof blew away in a storm in 2013 on the way to Abel Tasman National Park, 2013

PHOTOS CAMILLA RUTHERFORD **TEXT** KAYE MUELLER

Stoney Bay on Coromandel peninsula. Watching the sea. "Clara" takes a close look; this orange bus was the first Bulli in Bevan's fleet

Rakia river flows in New Zealand blue. Rakia Gorge was originally planned as a railway bridge. But no train ever passed over it. Thank goodness

Type
T2
Model year
1974
Built in
Germany
Power (kW/hp)
48/65
Engine (displacement)
1.8 l boxer
Colour
Orange and white
Special features
Clara's force of attraction

Name
Bevan Beattie
Year of birth
1969
Place of birth
Hastings, New Zealand
Place of residence
Silverdale, New Zealand
Profession
Businessman
Passions
Owner, enthusiast, and restorer of classic VWs. His wife says he's a hard-working jack-of-all trades

The Coromandel peninsula in the rear-view mirror.
The Beattie family: Ezra, Keralee, Josiah, Elise, Andrea, Bevan and little Priscilla with a 1966 Bulli at Classic Campers in Silverdale, a district of Auckland

number in the window, but no one answered when I called", he explains. "So I went back and waited for the owner, caught up with him at a Lotto shop, bought the '74 orange Bulli on the spot and named it Clara."

He continued working at his full-time job while establishing his Bulli rental business on the side. In 2005, he married his wife, Andrea, and today they have five children. Things then started to get a little tight, as Bevan explains: "My Bulli rental business was growing, I was still working full time and I also wanted to spend time with my family." So he quit his job, bought 2.5 hectares of land outside the city and built a big barn that is now the headquarters of Classic Campers.

"Most of our customers are couples between 30 and 40", Bevan says. "But the Bulli is also perfect for families: It has plenty of room, fits in any parking space, and doesn't consume much petrol. Dutch customers like the orange Clara, the Irish prefer the green bus and wedding parties want to have one in a colour that complements the bride's dress. Still, the most interesting stories are the ones where something goes wrong." For example, at the end of 2014, a major storm hit the west coast of New Zealand. Two very nervous customers from France called Bevan from there to inform him that "the roof flew off our VW!" Bevan told them to hurry up and go find it; a glass-fibre dome couldn't be too hard to locate. That wasn't going to happen, though, since a powerful gust of wind had tossed the roof into a raging river. Today, it's floating around somewhere in the Tasman Sea. The French customers hadn't fastened the roof properly." After that, we began providing detailed instructions on how to secure the pop-top roof", Bevan explains.

"There was another time when a project manager and a lawyer filled a drinking water tank with petrol. We've also had engine damage because people failed to check the oil or ignored the warning light that went on after the fan belt broke." That, of course, meant that the engine was no longer being cooled, and if the engine overheats, it is damaged beyond repair.

There are also stories of love: "Douglas, for example, is a professor from Canada who visits frequently and always rents the same TDI for three or four months. He studies sand flies. He loves the TDI, and it's likely we'll eventually name it after him. In fact, Bullis are like children: It's impossible to say which one you love the most. Each has its own quirks and personality that make it unique. In the end, you love them all." Fortunately, Bevan is willing to share his great love with others. ❤

ATM
ATUL
COLLECTI
ING
PAR
PH-2743640

MY BEST FRIEND

India. The epitome of the exotic. But what is exotic in India? Maybe Inderpal Singh Gill's travelling treasure, a savannah beige VW camper with a classic pop-up roof, which has been halfway around the world and yet still looks as though it has just come straight from the factory …

Inderpal's world has three dimensions. Family, of course. Then there are Ludhiana and Westfalia. Ludhiana is a city in the state of Punjab in northern India. It is located alongside the River Sutlej and the Grand Trunk Road, the region's major highway and has a population of 1.6 million. More bicycles and tractors are built here than anywhere else in India, and its agricultural sciences university is the largest of its kind in all of Asia. Another famous establishment is the Christian Medical College, the continent's first medical training institution for women.

Westfalia is a coachbuilder from Germany, known for its motor homes and among classic car fans for its VW campers with the classic pop-up roof. Such vehicles were built half a century ago and one of them, from 1969, now stands in Inderpal's garage. In elegant savannah beige, all spick and span.

Ludhiana by day and night. Despite its savannah beige colour, Inderpal's friendly Bulli stands out on the permanently congested roads

"My VW is my best friend", says the 38-year-old farmer and businessman, whose full name is Inderpal Singh Gill. "Things are hectic during the week, and my work eats up all of my time. But I spend my free days with the family, my wife and children – and my camper van."

A magnificent sight, outside and inside. Even after decades of use, the seat upholstery looks perfect. Now, however, Inderpal wants to show me his world and invites me to take a spin with him through Ludhiana with his best friend, the "Bulli".

Ludhiana – India's granary

It's a city where everything is both on the move and at a standstill: rickshaws and SUVs, buses and carts, bicycles and lorries tangle together into an almighty chaos, with the air filled with exotic fragrances one moment and horrible odours the next. We pass endless rows of shops with decorative bangles and bracelets in a thousand and one colours. Right next door they're selling hexagonal head nuts and bolts. Young girls are looking for silk ribbons for their jet-black hair. Here a shopkeeper checks a delivery of cables, over there an old farmer carts his harvest to the corn market, and in the midst of it all a young man whooshes through the hustle and bustle on his Harley Davidson without a care in the world. Totally exotic to European eyes. But to Indian eyes the only thing exotic here is the beige Bulli.

Inderpal's eyes gleam with pride. "We first saw the VW in 1999 and fell in love with it straightaway. I came to an agreement with the owners and we brought the car home in 2006." What the camper van may have experienced in the intervening seven years, Inderpal doesn't know. In any case, the Bulli came to him in excellent condition: "There were just a couple of things

PHOTOS BENGT STILLER **TEXT** SHIBANI VARMA

It's a man's world: the male headgear, the dastar, is different from the Arab turban. It is tied every morning and covers the men's long hair. Inderpal demonstrates it for us with the help of his daughter Kushi

that needed to be done. The logo was missing at the front and the glove compartment had no lid. Otherwise the vehicle appeared perfect!"

The camper came with two spare wheels. All of the tyres were "Made in Canada". "Even the number plate from British Columbia, dated 1979, was still in the car", recalls Inderpal, sitting at the wheel, which in German style is on the left. It's impressive how he manoeuvres the vehicle through the Indian street jungle, sitting on the wrong side and chatting casually at the same time: "The vehicle was flawless in every aspect. Nobody had tinkered with either the engine or the bodywork. And the Westfalia interior wood trim is simply unbeatable in its beautiful detail! The camper stood there as if it had just rolled out of the factory!"

That alone would certainly be remarkable for a VW veteran well over 30 years old. It gets even more astonishing when Inderpal alludes to its history: "It has been around the British Isles, along the coast of Italy and down the country lanes of Bavaria. It has crossed the wide open spaces of the USA, withstood the cold of Bulgaria and wound its way through the narrow confines of historic Greek cities. The Bulli has been on the shores of the Mediterranean in Syria and travelled to both Iraq and the spice markets of Afghanistan." From there, it made its way via the bumpy roads of Pakistan to India, all the way here to Ludhiana.

Inderpal steers the Bulli through the chaos of the city in a confident, matter-of-fact fashion, as if we were on a Sunday afternoon drive in the country. His passion for classic vehicles has a long tradition in the family. "When I was born, we had two classic Mercedes, a W 110 with tail fins and a W 115, the classic /8." My grandfather used to drive one and my father the other. "We still have the 115 at home." In fact, there's an imposing collection of classic cars in his garage: several Beetles, for example, classic Mercedes and a smart Italian Lambretta scooter.

Leisurely outings and historic rallies

The traffic around us has thinned out now. Gradually, an idyllic country scene stretches out before us. Inderpal seems to feel more at home now. Big city hustle and stress are not his thing. That's also why he lives with his family and the Bulli not in the city centre, but 20 kilometres out of town.

"It was a conscious decision. Living in the city wears you down. You no longer have any time or energy to indulge your passion." Passion. That's Inderpal's elixir of life. „A life without passion is like a body without a soul. So I drive the 20 kilometres to work. And in the evening, I come back to a peaceful environment, go into my garage and work on my cars. I enjoy the lovely life and the clean air out here."

The flapping curtains of the opened rear side windows and the scent of ripening fields and old wood that wafts in create a nostalgic allure. Inderpal and his Bulli are in their element here. He keeps his treasure in

Sandwiched between concrete barriers: this is where old and new India meet. In his Bulli, Inderpal stays calm in the traffic

25

Type
T2
Model year
1969
Built in
Germany
Power (kW/hp)
35/47
Engine (displacement)
1.6 l boxer
Colour
Savannah beige
Special features
Unrestored and yet has been around the whole world

Name
Inderpal Singh Gill
Year of birth
1977
Place of birth
Ludhiana, India
Place of residence
Ludhiana, India
Profession
Entrepreneur and farmer
Passions
Classic automobiles – and especially Volkswagens

All-round passion. In addition to his Bulli, Inderpal has a penchant for various air-cooled VW Beetles

good shape with regular excursions and ensures that only spare parts "Made in Germany" ever get installed in the Bulli. "Sometimes I go camping with friends with a few beers and a barbecue, or we have picnics with the family. My children like sitting down by the river, cooking. But whatever we do, the camper is always there. Occasionally, I do long trips on my own. Then I have peace and quiet for inner contemplation, and I feel at one with the surroundings. That is pure bliss."

To Inderpal, romantic yearning and technical know-how aren't incompatible. The man is an encyclopaedia of cultivated classic car knowledge that extends from the year a car was made to the years of production of individual models and even details such as which original bumper goes with which chassis number.

Classic VWs are part of Inderpal's world. And in the world of classic Volkswagens his is a recognised name. In Punjab, his VW camper van and his Beetles are great attractions at historic rallies and exhibitions. The rallies are generally in aid of charitable causes. Inderpal believes: "Growth is only genuine if it means growth for all. It is important to me to be able to give something back to society by taking my VWs to the rallies. Plus, it adds to the history of the cars."

He also enjoys driving to VW shows in Germany. "I meet people there who have the same interests. These shows are a great platform for exchanging restoration ideas, techniques and experiences." And here he adds a sentence that is characteristic of him: "Travel helps me to gain new perspectives."

There is one destination that Inderpal still hopes to enter into his Bulli's logbook: Leh in the Indian state of Jammu and Kashmir, the former capital of Ladakh. Situated 3,500 metres above sea level. Not a trip for the faint-hearted, but that is precisely what appeals to Inderpal Singh Gill. Not out of bravado, but for a much deeper reason. Almost reverentially he utters the word that says it all about him, the camper van and India: "Soul".

Happiness under the tilting sunroof. Kushi means happiness. The 7 year old is Inderpal's pride and joy. Together with the rest of the family they often enjoy going on picnic outings

美华
洗车美容
洗车补胎
羊杂汤
山东

THE FORBIDDEN T3

Hardly anyone rides bicycles any more in Beijing. A dignified saloon or an SUV stand for prosperity there and are status symbols. Bai Ning, however, drives a T3 van

There are over 100 Chinese carmakers. Any self-respecting Chinese province has to be able to show off at least one factory that manufactures cars. Around 40 years ago, things still looked very different. Then there was actually only the 'Red Flag' state limo, of which only limited numbers were built. The marque still exists today. But apart from a few party officials, nobody ever drove one, nor do they drive them today. Something, however, that has been much more enduring is Volkswagen's involvement in China. The company has had production sites in the country since as far back as the 1980s. The Volkswagen Santana became the epitome of mobility. For the young Bai Ning from Beijing, VW was 'his' dream make of car. Oddly, however, it was the VW van that fascinated him and not the relatively prestigious Santana.

As children of the '80s, he and the T3 are effectively part of the same generation. The panel vans of that time are the first vehicles that he can remember. "Basically, the T3 represents the 1980s style of design", remarks Bai Ning. "The straight lines along the side, the rectangular windows without any frills and the small, round headlights have a classic, simple look. I liked that a lot about the T3." And so the idea took root in him to buy such a vehicle himself. After many years of looking, he then finally found one: a consulate had imported a Bulli to China 20 years before, then sold it after 15 years of service. The owner before Bai Ning had the once silvery grey van repainted in its current yellow. He loved his T3 above everything and looked after it meticulously. Then along came Bai Ning, asked if it was for sale, persisted and eventually got the Bulli, the vehicle of his dreams. "The T3 still has a few defects, but I'll gradually bring it up to scratch and give it a new shine", he promises.

Built in 1988, the VW has an automatic gearbox and electric windows. From today's perspective, that was a top specification for the times. "The water-cooled flat engine being in the back makes it very quiet to drive", says Bai Ning. On bumpy roads, the Bulli copes very well and plays its trump card: long spring travel. "That feels good. Every journey in my T3 is an unforgettable experience." And he hopes that one day he and his family will be able to participate in classic car excursions across the globe with the Bulli. Bai Ning has goals. He will achieve them – with his VW T3.

Type T3
Model year 1988
Engine 2.1 l WBX, 82 kW / 112 hp
Special features Former consulate vehicle

Name Bai Ning
Year of birth 1985
Passions Classic cars, VW bus, the 1980s

zǎoshanghǎo – good morning! The T3 takes a cold outdoor shower. Bai Ning takes it calmly

TS T 34H

FLIGHT OF THE PHOENIX

The "Fokker" bus was all that rose from the ashes of his workshop. But Andreas Lerch didn't give up

Andreas Lerch will never forget the 24th of August 2014. A black day. Whilst he was enjoying riding his motorbike in South Tyrol on a Sunday in late summer, he was completely unaware that a fire was destroying his livelihood in Traundorf, Bavaria. Along with his vintage cars, spare parts and tools, the fire also destroyed his car for everyday use and several classic cars belonging to friends. Fortunately, no one was harmed. In those difficult times friends supported the father of two, organised a football game to raise money for him and raffled a Beetle for his benefit.

"At first, I didn't want to see the Fokker bus any more at all. It was in a friend's garage and his Bulli was in my workshop. If we hadn't swapped, his T1 wouldn't have been destroyed. That really upset me", Andreas Lerch says. His mechanic colleague turned down the offer to keep the Fokker bus. But now the classic car mechanic is looking to the future more optimistically again. "The bus is like a family member to me. Really! I've had every screw and bolt in my hand. And we also have a lot of fun together. When I'm on the road with my children, Laura and Lukas, with the Safari windows open – they never whine or ask if we are there yet or say that they are bored. And they don't want to go on holiday in a normal car at all anymore." That's why it's important to find an engine again for the 1955 bus. "It certainly won't be such a wild engine as the three-litre powerplant that was unfortunately lost in the fire, but it will definitely have more power than the production version!", says the man who has made Bavaria his home and has a passion for fitting tuned engines. "50 hp might be enough where the roads are flat. But here? There are mountains everywhere. And with power of just 44 hp or less working at the rear, my passion for the early Transporters would soon fade." What's more, such a slow pace would not be in keeping with the Fokker emblem at all – a reference to the aeroplane manufacturer that was initially German and later Dutch.

Andreas goes into raptures: "I like panel vans best of all. I really like the lines. Everything closed except the four windows at the front. This body shape in particular gives you a feeling of security. It can't be compared with anything else." Except perhaps the friends who picked him up again after the blow of fate. Thanks to them he has the courage to go on – because flames cannot destroy love and friendship.

Andreas Lerch isn't an aeroplane mechanic.
His métier is classic cars, especially those from Volkswagen. Most of all he enjoys working on first generation VW buses because "their faces quite simply move me!"

Type T1
Model year 1955
Engine Engine Currently none, previously 3.0 l boxer, over 150 kW/200 hp
Special features Too many to list

Name Andreas Lerch
Year of birth 1977
Passions Classic cars with powerful engines

PHOTOS STEPHAN REPKE **TEXT** GEORG OTTO

STEPPENWOLF

Be it Africa, Mongolia or a domestic beech forest, Andreas Lidl from Kamp-Lintfort regularly answers the call of the great outdoors. When he's on his own, there's only one thing he has to have with him: his T3 Syncro

At home yet far away. Where is he right now? With the young Massai who wanted to marry him? Or the bull elephant who had an eye on him?

Up we go!
Thanks to a special spring construction – "Andreas springs" – the overhang angle of the T3 Syncro has once again become much larger

He could make slide show evenings into slide show years – we would never shut our eyes. Andreas Lidl's experiences from the Far East and Africa would place themselves between our eyelids like matchsticks. His photographic archive would in essence contain enough material to fill Bulli Love on its own. However, as we have given his picturesque images a special place in the gallery, we would like to tell you a bit more about this rather shy person behind the steering wheel of the 1988 T3 Syncro. As even a president was unable to draw this man out of his shell. He had wanted Lidl to appear as a guest speaker at the Mongolian embassy in Berlin and to give an account of his travel experiences. However, Lidl, from Kamp-Lintfort in the Rhineland, turned the Mongolian statesman down – grand shows are not his style. The president should have spoken to him about his favourite vehicle, as that is guaranteed to make his eyes light up! Lights on: Lidl's first love was a T3 from 1981. It had a power output of 50 hp and was atlas white. However, that Bulli had to manage without an all-wheel drive system. With two-wheel drive, he set off with son Manuel for North Cape. Despite breath-taking scenery, it soon bothered Lidl that he was not able to take the routes that he, and above all his son, longed to take. Therefore, when the two returned, next on the agenda was a prompt visit to the nearest Volkswagen dealer. Lidl wanted to take a good look at the Syncro model of the VW Bulli. After a half-hour test drive and pushing it to its physical limits in the local brushwood, the reading on Andreas Lidl's personal Richter scale was sky high. His heart had been captured by his first T3 Syncro. In Riyadh yellow.

Now at last he could conquer the world, and he didn't want to wait a second longer. After a year of planning and eight semesters of Arabic at the community education centre, Libya beckoned. In the sand hills of the Sahara, the ride was a constant up and down. "Once we even turned

PHOTOS THEODOR BARTH **TEXT** BASTIAN FUHRMANN

Type	**Name**
T3	Andreas Lidl
Model year	**Year of birth**
1988	1956
Built in	**Place of birth**
Germany & Austria	Duisburg, Germany
Power (kW/hp)	**Place of residence**
88/120	Kamp-Lintfort, Germany
Engine (displacement)	**Profession**
1.9 l in-line TDI	Electrician
Colour	**Passions**
Timor beige	Outdoorsman – always true to the Chinese proverb: "Sheep run in herds. Eagles fly alone."
Special features	
Combination of a potent SUV and a homey travel trailer	

over and ended up on the roof!" With the help of a following vehicle and a cable winch, Syncro and Lidl set their hair straight again. "The Libyan dunes are treacherous! But that was a mistake," he adds, "that I won't be making a second time!" Using his hand, Lidl draws for us the ideal line for driving up a dune into the desert sky. "Always at a diagonal angle to the wave of sand and to its crest. Getting slower at the top, so that you can see the sand crater behind it!" Understood!

"The Sahara will do with you what it will. Despite looking monotonous, if you're not constantly alert, you can be dead the next second." After several days and nights in the desert, the lover of solitude was then drawn back again towards civilisation.

"After an uneventful three-day drive without seeing a single soul, it does eventually get a bit odd!" Even for solitude lover Lidl, who in moments alone happily sits on the rear bench seat and reflects quietly for hours. The only human being in his vicinity: a photograph of his son stuck to the interior trim.

Today a work machine, tomorrow a camper.
"The T3 is a real workhorse." No moaning and groaning – or at least only occasionally as it moves through the undergrowth

Close your eyes and keep on driving

When the radio does nothing but hiss, and you can no longer bear to hear your own singing, for this Bulli man there's still the option of closing his eyes. While he's driving! A game for passing the time on straight runs. First for ten seconds, then thirty – then for a whole minute. Back in civilisation in the form of a camp site – life and activity. And danger from a whole different direction.

Someone stole his yellow T3 Syncro, which has never turned up again since. He had to travel back by plane. The end? No. He didn't want to be stopped by human cunning, and so the traveller bought a T3 Syncro in Timor beige from a friend in 2001. The vehicle of his dreams, built in 1988 (in Hannover and Graz), has a 1.9-litre turbo-diesel engine, 120 hp, solar cells, two helpful differential locks and "Andreas Springs".

The springs, which he named after himself, were designed and patented by Lidl, a trained electrical engineer. They are his bit of extra ground clearance, which helps him just as much when getting wood for his grandmother's stove as they do on the steep sand dunes of some delta of his personal freedom.

In 2009, he embarked on a journey to Mongolia, the adventures of which the Syncro fan recorded for his son Manuel in a book. It shows how Lidl crossed Russia, Siberia and the Gobi Desert and ultimately reached Mongolia via Kazakhstan. An incomparable adventure.

The book for his son closes with three words. We quote them here and offer them in tribute to his T3 Syncro: With love, Andreas.

Moss green meets timor beige. The solar cells charge up in the midday Lintfort sun. In the Sahara they supply enough electricity to power the integrated coolbox – "A cold beer whilst travelling is a must!"

ANTIK
Konviktstraße
Luitpold Baue
WAREN

THE CALL OF FREEDOM

Schauinsland mountain rises high above the Black Forest just outside of Freiburg. The Vosges Mountains and even the Swiss Alps can be seen from here. The peak is a popular jump-off point for paragliders. Peter Vaas, however, sometimes just likes to circle around above the small red-white dot on the ground – the vehicle that brought him up the mountain: his beloved red-and-white T1

Freiburg im Breisgau – the Old Town. Tyres can be heard softly tapping on old cobblestones. A 1966 tornado-red/cream-white T1 gently comes to a stop in front of us, right opposite Zum Roten Bären, Germany's oldest inn.

Peter Vaas, a man as versatile as a Swiss army knife, jumps out of the bus. There's nothing this man cannot do – nothing he hasn't done, or at least hasn't tried to do once. Indeed, he's led a many faceted life, but one thing has remained constant: his Bulli. As he drives us up one of the most beautiful serpentine roads in the Black Forest, all the way up to Schauinsland mountain, Vaas, whom his friends call "Pit", tells us about his interests and his vehicle. The vehicle, a Bulli T1, carries the equipment on its roof that brings Vaas to the mountain on a regular basis: a paraglider.

When our photographer, Theo, and I entered the bus, the springs in the seat upholstery immediately gave way, causing us to sink down deep: the Bulli feeling had begun. Vaas's first encounter with a Volkswagen Type 2 dates back to his childhood, he tells us as we rattle through the Swabian Gate in Freiburg.

"My grandfather had a T1 bus that he used as a delivery vehicle for our printing business", Vaas explains. "I thought it was fantastic and would sit in the passenger seat as often as possible."

At some point, Grandpa Vaas wanted his grandson to take over the family business. Vaas was only 22 at the time. Although he was a trained book and offset printer himself, he couldn't imagine spending his days in an office and assuming responsibility for 120 employees at that age – even with that T1 standing enticingly in the company parking lot.

A roundabout route to the T1

Instead, he wished to – had to – pursue his powerful desire to simply be free. He wasn't sure what form this freedom would take; he assumed he would find out in due course. One of the first things he did was design an indoor rock climbing centre, which he then built with friends. The facility in Satteldorf became Germany's first indoor playground for adults and even attracted professionals such as the extreme rock climber Stefan Glowacz, who would later offer courses at the centre.

Vaas then switched from rock climbing to motorbiking and entered various endurance and trial competitions. "It was all rather extreme in some ways", he says. "With a lot of these guys, it was like they were battling with knives in their teeth. I wasn't interested in anything that brutal, so I finally decided to do four-hour races."

The wings of the fixed wing paraglider and the friendly face of the T1 and its owner appear. Who said anything about antique?!

We are about to reach Schauinsland road, a former serpentine which the T1 climbs effortlessly up to 1280 metres. The skies open – our ears pop

During this time, he repaired heavy motorbikes to finance his racing hobby. "But after I started my own family, it all became too much for me", Vaas explains.

We now pass the station in the valley for the Schauinsland cable car. Vaas points to a small field of grass behind the local playground. This is where he and other paragliders land safely – with a group of little fans cheering them on behind a fence. The plan is for us to pick him up there later – after he's done circling around in the sky. But first we need to get to the top of the mountain that looms ahead. Despite its relatively weak 54-hp engine, the T1 bravely handles the curves on the way up and makes the climb with no apparent problems. Our ears begin to pop, but the rumbling of the 1500 engine seems to amplify even more inside our bodies.

Vaas now sounds like he's talking through a giant wad of cotton as he explains to us how he went from motorcycles to paragliders. The explanation is quite simple: "by way of the water", he says, and then describes how he once went whitewater rafting when he was 12 and the river he was on seemed to stand still for a moment. It was then that young Pit saw a paraglider artfully floating and turning in the sky. "He seemed to possess such lightness and style as he flew above us – it was simply splendid!", the Bulli aficionado recalls. "I wanted to do that too at some point."

It seems quite logical in retrospect: from land to water and then to the air! Vaas's to-do list thus came to include obtaining a paragliding

Tales of old-town Freiburg. An admiring glance at the former Merdinger Fire Brigade vehicle.
For Peter Vaas' Bulli the fire still burns bright – especially here in Baden-Wuerttemberg

Pit looks up to the skies. When the thermal conditions are right and there is an east wind blowing it's time to head up the mountain. Spread out the glider. Take a running jump. Admire the Bulli from above

licence from the DHV hang-gliding and paragliding association in Germany. Later, of course, he got himself a licence as a paragliding instructor as well. A short time after that, Vaas went to work for an air-taxi company in Austria that brought people down from mountains into the valley.

He also taught gliding on the side – one of his students was actually an 80-year-old nun. "She probably wanted to get a little closer to God", Vaas says with a twinkle in his eye as he guides the Bulli through a hairpin turn and shifts down a gear.

Downshifting – now that's something Vaas also knows about: despite his love of flying, he avoids getting too close to God too fast. "A couple of years ago, the doctors said I might have suffered a heart attack, so I take things easier now", he says. "I try to give myself and my employees the day off on Friday", he adds. Here, Vaas is referring to the people who work at the small gardening company he now operates. "I want to be around to watch my kids grow up, after all", he explains, "so I usually only fly on Fridays. The Bulli is a big help – it's the most beautiful form of taking it slowly!"

It was Vaas's wife, Susanne, who suggested buying a Bulli. Vaas himself had previously had only bad memories of the Type 2 Volkswagen.

That's because a night drive he took with such a bus when he was 20 nearly cost him his life: he was driving a T2 one wintry evening when he lost control of the vehicle in a sharp turn and ended up on a frozen lake. Fortunately, a thick sheet of ice kept him and the Bulli from sinking completely; they were only half submerged in the lake. Vaas escaped the icy trap by climbing out of the rear hatch. He walked to the nearest farm house, woke up the farmer and asked for help. In his mind, he could see the T2 sinking deeper and deeper into the lake. Luckily, another farmer had the equipment needed to safely pull the Bulli out of the water.

Vaas and his Bulli were then reunited on dry ground. The T2 actually started up – but the accelerator cable was frozen at full throttle. Vaas then tried to turn the ignition key to the off position at the first traffic light he came to, but it didn't work. He was determined to go home, however, and he probably got there faster than ever before, as he sped to his house and then pulled on the parking brake to bring the T2 to a halt in the gravel in his front yard.

The next day, he worked on the lines that had completely frozen overnight. He and his sister tried to salvage what they could using a hair dryer. All their efforts were in vain, however – the damage was simply too great and the T2 had to be junked.

Come let us fly!

Vaas met his wife, Susanne, 11 years later. Their love for one another also led to a sort of rebirth of Vaas's long-forgotten passion for Bullis. It turned out that Susanne had owned a Samba when she was a student in Freiburg. She later had to get rid of it for financial reasons, but she still missed it a lot. "Don't be sad; you'll get another one sometime", Vaas told his wife. He then promised to find her a Bulli.

As Vaas tells his story, we keep climbing higher and higher up the serpentine road in the Black Forest. The reflecting directional arrows at the edge of the road have a pale appearance. It's getting colder, and we now see remnants of the snow that fell here in the winter. Vaas continues with his story as the small toy ducks suspended from his rearview mirror sway back and forth.

He comes to the part where they finally found their Bulli. After they got married, they went camping but forgot to take a gas canister with them.

The escape route from the ice led outside through the tailgate. Half his T2 was still stuck in the lake. Hours later the T2 headed home at full speed with a frozen throttle cable

Type
T1
Model year
1966
Built in
Germany
Power (kW/hp)
32/44
Engine (displacement)
1.5 l boxer
Colour
Cream white,
tornado-red
Special features
Former firefighting vehicle

Name
Pit Vaas
Year of birth
1964
Place of birth
Crailsheim, Germany
Place of residence
Kirchzarten, Germany
Passions
Flying, motorcycle riding,
VW Bulli

That's how they found their T1 Bulli, whose friendly face stared out at them from a solitary parking space at the camping equipment store.

It had a faded tornado-red/cream-white body colour, and grass was growing out of its interior. Not the type that is smoked, but the type that is pulled as weeds.

"Apparently, the Bulli had been through a lot", Vaas explains. "You wouldn't have known it at first, but it had previously been used by the fire department. Two students later bought it from the fire brigade and converted it. They put the wrong roof on it, however, so I had it replaced with an original roof. Besides that, I only customised the interior to make it more practical for me and my family – basically, I turned the Bulli into a motor home." He did all of this with his own two hands, of course.

Vaas now slows down the Bulli and brings it to a stop in a parking area. This is where the T1 climb ends and the flying begins. Vaas puts on his backpack and steps over a guardrail. A few metres further we come to an open space between the trees. Vaas spreads out his glider, quickly checks the thermal updraft and fastens himself into the loops. His rate of descent indicator beeps softly. The atmospheric inversion he was hoping for has arrived: an easterly wind. He's now ready to take off.

He runs up ahead to fill the glider's wings with enough air and then jumps straight off the cliff. For a second we think he's in a dive, but just as quickly he reappears in front of us. He uses the updraft to fly in circles. As we watch him, we wonder whether it's the flying that marks the highlight of his trip up the mountain with his Bulli – the backdrop of mountains such as Mont Blanc in the distance. Or perhaps the highlight is afterwards, when he starts the T1's engine, puts his hands on the old steering wheel and heads home down the mountain to his family – to his wife, Susanne, his children, Jannis and Anna and his dogs, Lilly and Paula.

The rate of descent indicator beeps, the tornado-red dot is waiting

Vaas turn his circles way up in the sky, the tornado-red dot is waiting. He could go anywhere, but he chooses to circle above the small tornado-red/cream-white dot on the ground. It's as if he's connected to it by an invisible rope. Pit Vaas and his Bulli: together forever. ❤

Black Forest gang – Bulli family Vaas at home in Germany's sunniest town. And then there's this Bulli. Life can be hard sometimes!

VOLKSWAGEN
FR V 1966H

ON THE ROAD FOR A THOUSAND LIFETIMES

When Ben Jamin set out from Chile to travel to Alaska, he wanted to experience adventure. But he never would have guessed that he would have so many. Indian chiefs in the Amazon rainforest, surfing paradises – and a few repair shops awaited him. Hasta Alaska!

Was it luck, was it a blessing or was I simply a crazy person? Call it what you will. The day when I set out from Chile to drive to Alaska and find adventure in my 1992 VW Kombi was a dream come true. I clearly understood that a lot of other people dream this dream as well, and it is my passion to share the experience of this adventure with a lot of people. That is why I open the doors of my small hippie home to anyone who wants to share the experience with me. Those who cannot join me can still drop by to see "Kombi Life" on my YouTube channel.

I want to inspire people to set out on their own expeditions – whether big or small. Travelling is the best school of life that a person could ever attend, and it is possible to cover very large distances on very little money.

Co'Pito (Spanish for "flake"), which is the name of my sometimes temperamental mobile carrier, is a VW Kombi from Brazilian production. It is a cross vehicle, if you will – a cross between the platform and technology of the T1 and the front end of the T2. Fans sometimes refer to it as a T1.5. At some point, after innumerable engine overhauls, I fitted it with a fuel-injection engine from a Mexican Beetle from 2001 production.

So far, more than 100 people have travelled in our vehicle on the slow journey to the far north, including two chickens and a very stubborn little dog from Peru. Each day brings enough adventure for a lifetime. I have haggled with Indian chiefs in the Amazon region to secure our passage, and negotiated with hundreds of corrupt officials (not once having to pull out bribe money). Along the way, I touched the lives of many people, saved a few, almost lost my life a couple of times – and I never felt more alive than during this time!

A road trip like ours – over an entire continent – takes any classic car to its limits, and I must admit that occasionally I demanded a little too much from my Volkswagen. Often enough, there were ten people aboard the bus, and the streets were among the worst on the planet. So it is no wonder that our bus needed a lot of my tender loving care and my time to stay healthy and run well. Sometimes our Volkswagen actually had a cunning way of getting me to immerse myself deeper in foreign cultures – namely when it did not want to run any more. People often ask me why I didn't buy a more suitable car for my adventure. After ten engine

This isn't a postcard, it's a stop.
Ben Jamin and Co'Pito happy together

TEXT AND PHOTOS BEN JAMIN

A photo shoot on the beach with Ben's bus as a backdrop. Every day brings new impressions. Time and again the adventurer is moved by the friendliness of the people. And the journey hasn't finished yet. It's a long way to Alaska – and the bus doesn't seem to be in much of a hurry to get there

A Co'Pito is attractive. The surprise of the spontaneous bikini photo shoot is written all over the Volkswagen's face

Type
T1.5 Kombi
Model year
1992
Built in
Brazil
Power (kW/hp)
37/50
Engine (displacement)
1.6 l boxer (currently with fuel injection)
Colour
White
Special features
Tendency towards dramatic behaviour, doesn't like long journeys

Name
Ben Jamin
Year of birth
1981
Place of birth
Jersey Island
Place of residence
World citizen
Profession
Producer, director, author
Passions
Travels and adventure

A haircut on the beach. Then the right reading matter in the shady bus – a simple life yet much richer than a package holiday could ever be

overhauls and nearly twelve months of dedicated maintenance, I ask myself that as well. On the other hand, there is no other vehicle in the world that can magically elicit smiles on so many faces. The Kombi brings out another side in all of us, and it unites people in an inexplicable way. I have experienced it myself – from a businessman in his 100,000 dollar ride on Hollywood Boulevard as well as impoverished children of the natives on the high plains of the Andes. It does not matter who you are or where you come from – you simply have to turn around and look back at the Kombi, because it appeals to your heart.

The key to the family

In retrospect, it feels like I did not just buy a car in Chile three years ago; rather, I became a member of a family. On my trip, VW enthusiasts showed a lot of love. They helped me up close and from afar when the car's mechanical equipment refused to cooperate any more, and many actually invited me into their homes. In the USA, VW fanatics got together and sent me manuals and parts, which I used to rebuild the engine and bring the VW back to life after a piston seized up in Costa Rica. The support of all of these people is the only reason that my Kombi life project stayed alive, which is why I will be giving my VW (my home, my car and the only thing in the world that is of any value to me) to one of my fans at the end of this adventure.

Not long ago, I experienced another wonderful surprise early one morning on a beach in northern Baja California, in Mexico. The Volkswagen Club of Tijuana, this city that is close to the USA border, showed up in full force and brought dog food for my puppy "Alaska" (named after our destination). They also brought blankets so that we wouldn't get cold on our journey northward.

They insisted on checking our VW thoroughly in their shop, and this inspection revealed a couple of serious problems. A cylinder head was no longer seated firmly, and the front wheels were nearly ready to come off the axle. For two weeks, we worked side-by-side, searched for parts in Tijuana and simply built those parts that we could not locate. Every day, I woke up to

Dramatic clouds, absolute solitude. Bad weather can't threaten with Co'Pito as a driving companion

coffee in the shop, and every day I ate three homemade meals – Mexican specialities. I really should have stayed in Tijuana if only for the food.

It was an unforgettable experience to immerse myself in the life of Tijuana and to enjoy the very generous hospitality there. When I reluctantly said my final "goodbyes", no one wanted to take my money. Just the opposite – they gave me all the money back that I paid for parts. I also got a jacket with the embroidered name of their club president and an invitation to even be a member of their Volkswagen family.

The new friends would love to come along, but...

During my two short weeks in Tijuana, many of my new friends expressed a desire to travel with me. I would have gladly fulfilled their wishes, but not everyone can actually pull up roots and go. I am very lucky. Not only was I able to undertake this epic trip, but I could also share the experience with millions of people at home thanks to new technologies.

When I set out, my vision was to dream this dream with a lot of people. Now we are sharing a life of adventure. Regardless of what comes or how tough it is. I will see this adventure to its conclusion and arrive in Alaska. ❤

The engine has quit working almost once a month so far. Ben is becoming an expert mechanic. But that wasn't his plan

HAPPINESS IS A HOT VW

Looking at Helge Ohmes, you would think he rides a Harley or drives a hot rod. Yet he drives a VW camper. With little flowers!

Helge Ohmes and his 1969 T2a go together like Hamburg and St. Michael's. The two only ever appear as a team. Well over six-foot tall, Helge is one of the best-known characters in the VW scene for air-cooled engines in the motherland of the VW camper – and internationally as well. That's due not least to the fact that he has made his hobby into his job. One of his jobs is visitor guide for the "Prototyp" car museum in Hamburg, and his other job is as an independent parts dealer – specialising, of course, in everything related to old Volkswagens. Whenever he sets up his sales booth at any of the countless meets in Europe, the 1969 T2a is always there with him. As a towing vehicle for the trailer. As a sleeping car. And as an eye-catcher! "People simply find the VW camper van endearing", says the 45-year-old. And yet initially he hadn't wanted a camper at all. His Bulli crossed his path more or less by chance – at a petrol station in Hamburg where he had stopped to refuel his everyday Beetle back in 1997. The petrol station attendant spoke to him and asked if he'd be interested in a camper van he had inherited. Not exactly enthused, the VW fan scheduled a day and time to look at it. And that changed everything! "Besides its fantastic basic condition, above all I fell for its smile. Unlike the T2b, the round bumpers and indicator lights positioned below it make the face of the early T2a models much more harmonious, with much more of a classic car look. And they simply drive better than the Bullis with a split windscreen." For a while, the vehicle remained in its purchased condition. Then came the restoration. His wife, Monika, made a plea for camping fixtures, while Helge himself made a case for a more powerful engine. "If, as we do, you frequently drive long distances for a short weekend meet, there's no harm having a bit more horsepower." And with a cheery wink he adds: "With an engine like that you can also leave other drivers wondering in amazement as you pull away from the traffic lights!"

So there's nothing left to be desired. Or is there? "To go on a proper holiday." That's something the pair have not managed yet with their camper van. "Our dream would be to drive the T2 up into the Alps. And then perhaps to carry on towards the Mediterranean." Until that time comes, the Ohmes family regularly uses the 1969 Bulli for shorter excursions, because "to me the camper van is pure joie de vivre! Or as the late Swedish racing driver Sven-Olof Brorsson always used to say: 'Happiness is a hot VW'!"

Type T2
Model year 1969
Engine 2.0 l boxer, 90 kW / 122 hp
Special features Good for long trips despite customised engine

Name Helge Ohmes
Year of birth 1970
Passions Classic VW, Porsche & US cars, classic car racing, customised & hot rod culture

Don't worry, he only wants to tell you something!
Helge Ohmes works as a visitor guide at the "Prototyp" car museum. Few people know more about air-cooled VWs or Porsches

PHOTOS STEFAN BAU, JAN BÜRGERMEISTER **TEXT** GEORG OTTO

MONSIEUR X

When he was just 22 years old, he caused a sensation in the Formula 1 scene behind the wheel of a Formula 2 car. In his day, Jacky Ickx was considered a daring, bold and relentless race car driver. These days, Ickx, who hails from Belgium, is retired and has learned to enjoy the blessings of a slower lifestyle by travelling with his wife through Africa – in a T5 California

It's quite a hectic day in Brussels when we meet up with Jacky Ickx in the Belgian capital. It's one of those days when all the heads of state and government in Europe have come to Brussels for a conference. A large part of the downtown area has been cordoned off – and all of the other roads are hopelessly congested. Certainly not the most favourable circumstances for a photo shoot.

Jacky Ickx, however, is not only extremely charming today; he also appears to be extremely calm. Totally relaxed, you could say. That's a bit hard to believe when you look back on his career – a continual story of life in the fast lane. At the tender age of 18, Ickx won the Belgian motorcycle trials title on a Zündapp motorbike. He then switched to four wheels, to a Formula 2 car, and shocked the elite class of the Formula 1 series by posting a qualifying time at the 1967 German Grand Prix that was better than those of all but two F1 drivers, this despite the fact that his Formula 2 car had 200 hp less output than his F1 competitors. What followed was the incredible career of a man who was probably the most versatile race car driver of all time. Indeed, Ickx raced for the biggest teams in the Formula 1 series, finished second in the World Drivers' Championship twice, won eight F1 races, took first place six times in the legendary 24 Hours of Le Mans race, competed successfully in the Touring Car Championship and crowned his career as a rally driver with a victory in the Paris–Dakar Rally in 1983.

Ickx was exposed to cars at a very early age because his father was a motoring journalist who used to bring home interesting test vehicles on occasion. That's how young Jacky became acquainted with the Bulli. "My father took us to the Black Forest in the first model ever released", Ickx recalls. "What a test drive – it was really a lot of fun." His love for the Bulli has endured over the decades and all the different Bulli model series.

Hey Jacky! With a spring in his step, the forever young Belgian walks towards us; the parked T5 awaits him

Race car driver? It simply happened that way

It seems hard to imagine a former race car driver behind the wheel of a VW bus. But as Ickx points out: "Actually, I never wanted to be a race car driver; it just happened. My real dream was to become a gardener. Or a gamekeeper. Something to do with nature." The four-wheel drive bus is perfect for this: "Yes, and I also had it fitted with some special equipment – more powerful dampers, bigger wheels and protective plating for the floor", Ickx explains. "You definitely need such things for Africa." Not that it's his first time, though: Ickx's second wife, Khadja Nin, was born in Burundi, and they both enjoy taking frequent trips through the northern and western parts of the continent, to places like Morocco, Mauritania, Mali and Senegal. "Sometimes we stay for a month, sometimes for two weeks", Ickx explains. "In the first

Type
T5
Model year
2013
Built in
Germany
Power (kW/hp)
132/180
Engine (displacement)
2.0 l in-line
Colour
Reflex silver metallic
Special features
The perfect travel vehicle

Name
Jacky Ickx
Year of birth
1945
Place of birth
Brussels, Belgium
Place of residence
Ohain near Brussels and the whole world
Profession
Race car driver, world explorer
Passions
Driving fast and travelling slowly, preferably in Africa

A beautiful view. Ickx Multivan is always parked in its private boxer pit lane. Always ready whenever he is

bus we had, we travelled 60,000 kilometres off-road, and we never had a single problem." His current bus is his second one.

Neither the vehicle nor its driver have any problems dealing with the traffic in Brussels today. After Ickx manoeuvres out of one last traffic jam, the fascinating Atomium glistens up ahead at the centre of a broad and majestic boulevard. The structure, a unique and famous landmark of Brussels, was built for the World Expo that was held there in 1958. A symbol of the atomic age, it has the crystalline shape of an iron molecule – just 165 billion times bigger. The building never fails to impress.

Today, it's surrounded by numerous billboards that call attention to a major exhibition at the nearby Expo Palais that will celebrate two Belgian national heroes: Jacky Ickx, a giant behind the wheel of a car, and Eddie Merckx, a giant at the pedals of a racing bike, both of whom will turn 70 this year – and look much younger than their years.

To this day, the two also remain synonymous with an unbreakable will to win. Two alpha males, you could say. Merckx was referred to in his day as the "cannibal" because he battled to win every time, no matter how minor the stakes. Ickx, for his part, once gave a simple one-word answer when he was asked what drove and motivated him in his races: "winning". When asked whether the speed of racing gave him an adrenaline rush, he said with an unintentional play on words: "It's no big deal – you actually get used to the speed rather swiftly."

One can quickly get used to a slower pace as well, or at least Ickx can. "Anyone who drives a VW bus like mine is thinking green", he says. "You really have to love and enjoy nature. It's also great to be able to stop wherever you want and always have your bed with you." Ickx is on the road a lot, but does he consider Brussels home? At this, his eyes narrow. "You don't happen to be from the police or the tax office, do you?" he then asks with a smile. "It's a little complicated: I travel for five or six months, then spend a month in Brussels, and the rest of the time I stay in Monaco."

Ickx, or rather his family actually, lives in Brussels in a sterling neighbourhood south of the city's imposing Palace of Justice. An Italian restaurant on the corner serves as his second dining room. "It's actually my main dining room", Ickx says with a grin. He's converted the ground floor of his home into a garage – "otherwise I'd be drowning in parking tickets". The garage is large enough to easily accommodate his VW T5 California bus – when he's not driving it around Africa, that is – or as long as no one else in his family is using it. The bus is very popular in the Ickx family. "We have a sort of car-sharing thing going", Ickx explains, whereby most of the sharing is with his five children.

Africa is ever-present - one continent, one love

Ickx's great love, besides his wife, family and Bulli, is clear to see in his home, which is decorated with carvings, pictures, carpets and design patterns from Africa. It's like a different world, but it's not the only one here, as the coffee table in Ickx's living room is also piled high with comic books. This isn't surprising, given the fact that Brussels is the world capital for comics. One of the more popular comic book figures of the 1970s was Michel Vaillant, a good-looking and very successful race car driver. "I also appear in a couple of issues", says Ickx, "but I don't like the stories because Vaillant always wins the races and that's not fair." Ickx now tries to look angry and contrite as he gazes into one of the books, but he's not entirely convincing. As is often the case, a thin smile can be seen at the corners of his mouth. There's actually a lot of Jackie Ickx in Michel Vaillant, but there's also a lot more to Jackie Ickx than just motorsport. That's the part that comes out when he gets behind the wheel of his VW T5.

86
88

ADAC
F ZI 22H

HOPE SPRINGS VELVET GREEN

To help people - that's Michael Steinke's passion. Vans used for that purpose include his T1. Because in conflict situations, a T1 is a peace-maker and silent negotiator

Shots whistle through the air. Had they been too daring? In the fog, a T1 panel van had driven off the road. A nervous border guard fired into the fog, and missed. Fortunately. That was in the intense period of the war in the Balkans at the beginning of the 1990s. As the organiser of an entire troop of old Volkswagen vans, Michael Steinke was delivering relief supplies from Germany to the disintegrating Yugoslavia. "Bullis Bring Joy" was the name of the initiative, which has so far taken over 130 tonnes of donated materials into the war-torn region and other parts of Eastern Europe, in a total of 70 convoys to date. "With its endearing charm, the T1 tore down barriers and opened doors and hearts to us", says Steinke, whose matter-of-fact manner belies his very caring nature. Naturally blessed with organisational skills, the van expert saw the great need and collected clothing, medication and small items of furniture. Six classic T1 vans were on hand for the first tour to Romania – a charitable undertaking that battled with temperatures of as low as -22 degrees Celsius and with customs officials, who liked a bribe – especially in the form of German police insignia. Steinke learned that antibiotics were deemed contraband, so difficulties were a certainty. However, everything always ended well. "These days, we use T3, T4 and T5 vehicles for our trips. You can simply get more in them." Another of his matter-of-fact statements. But Frankfurt resident Steinke, who doesn't see himself as a "fan of classic vehicles, but more as a user of vans" has been rumbled. Anyone who has driven Volkswagen vans for almost four decades and knows more about the Bulli than any archive, appreciates more than mere practicality.

The spring-cushioned swivel seat from a T2 helps on long tours. Or to exit quickly to help his neighbour, a 92 year old lady, seconds after this photo was taken

World's largest collection of Bulli brochures and catalogues

When he was about 20, Michael Steinke became familiar with the advantages of a 1964 VW van. At that time, he was topping up his student pay as a driver and made deliveries to the customers of a greengrocer. His boss was impressed by his reliability. The reward: the soon-to-be teacher was allowed to have the company van for personal use on weekends.

And he got the bug. Michael Steinke saved up and bought his first van, a second-hand Type 22, in which he travelled to Florence in 1969. His companion on this journey was his girlfriend Christa, whom he married shortly after this romantic adventure and who to this day shares his enthusiasm for VW vans and campers. In Tuscany, incidentally, he gained brownie points with his sweetheart by reciting to her details of numerous churches and Renais-

Type
T1
Model year
1967
Built in
Germany
Power (kW/hp)
32/44
Engine (displacement)
1.5 l boxer
Colour
Velvet green,
pearl white
Special features
Westfalia Conference
version (SO22)

Name
Michael Steinke
Year of birth
1947
Place of birth
Lehe, Germany
Place of residence
Frankfurt am Main, Germany
Profession
School headmaster (retired)
Passions
Helping people,
using Bullis,
VW history

sance buildings as her personal tour guide – his school trip to Florence had paid off after all!

The couple's holiday premiere in their first new car also took them to Florence. The 1971 T2a with Westfalia Helsinki trim had cost 13,000 marks. Two thirds of this sum had been assiduously saved up. The young teacher borrowed the rest from friends, so great was his desire to own the camper. Many more followed. Some they kept for quite a long time. "Passenger car aberrations", as Michael Steinke called them, have happened on only three occasions so far. However, they only tolerated these cars temporarily until the new VW van, ordered but not delivered yet, was ready to collect from the dealer. Usually camping versions. His knowledge of all Volkswagen vans and campers grew, and Michael Steinke became a "guru" of the German VW camper van scene, a man who simply knows the score.

It was not by chance that for many years he was chairman of the largest association of Bulli drivers, the "BulliKartei". His collection of brochures, leaflets and documentation, especially on the early T1 and T2 models is probably one of the most comprehensive in the world. Over the years, Michael Steinke has written ten publications on the subject of VW vans and campers. He is a fan. But a matter-of-fact one. One who values practicality and to whom museum deification of Volkswagen vans is anathema. He meticulously added up how many kilometres he has driven behind the wheel of a VW van between 1969 and 2015 for us. The total is around 1.3 million, or the equivalent of about 33 times around the globe. You can't do that without love. Even if it comes in such a matter-of-fact guise. ❤

A family affair. Christa Steinke shares the bus passion and goes on trips as often as possible. The plastic mesh of the front passenger's seat supported the back of Steinke's father, back then in a 1960 Ford Taunus 12M

LIKE A FLOWER IN HIS HEART

On picturesque Lake Constance on the German-Swiss border, there is a man who daily uses a T2 Bulli as a "flower vehicle" – and has been doing so since 1968. It's not that he's failed to grow out of his hippie lifestyle; it's simply his job: Hans Hodapp is a florist with all his heart and soul. He's also a Bulli driver – and if that isn't "flower power", then what is?

Type
T2
Model year
1971
Built in
Germany
Power (kW/hp)
37/50 hp
Engine (displacement)
1.6 l boxer
Colour
Elm green
Special features
2 sliding doors, first owner, old Überlingen numberplate

Name
Hans-Peter Hodapp
Year of birth
1948
Place of birth
Überlingen, Germany
Place of residence
Frickingen, Germany
Profession
Gardener, now retired
Passions
Everything that blooms

It's 7 a.m. Suddenly everything turns green in front of us. We heard the "Roarin' Sixties" long before we could see anything. The elm-green panel van races around a curve towards the cathedral square in Überlingen. The sunflowers we see through the rear window are leaning forward and are happily bobbing up and down. We're happy, too, as this had to be the Bulli, which since 1968 has never had another owner except its first, and has never let that owner down either. The panel van rattles straight towards the market square and stops right next to the fountain and does its usual brief and gallant "knee bend". Looking at the Bulli, you'd think the fountain might be the proverbial fountain of youth. At the very least, the T2 is in remarkably good condition. Its driver, who of course is dressed in a matching green, steps out of the van to greet us in his own special way. "Good morning", he says, "but sorry, first I have to take care of my customers". The conscientious precision of a Swabian.

"No hurry", says Markus, who's familiar with the Swabian region (it's where he chose to live) and knows how to get along with the local crowd. However, one should never underestimate the Swabian capacity for enthusiasm. For example, shortly thereafter, a woman driving by stops her car, runs out and expresses her pleasure at seeing the obsolete numberplate on the Bulli bearing the abbreviation "ÜB" for "Überlingen". "ÜB – I never thought I'd ever see that again", she says, and departs. Hodapp – the flower enthusiast – pays attention to detail, and his unchanged numberplate, along with his Bulli, have made him somewhat of a local hero.

Now, the florist is all business, though, as he sets up a table and, to our great surprise, opens sliding doors on either side of the T2– a custom addition back in the day. Two wooden panels, one above the other, serve as shelves for plants and make loading and unloading a cinch for Hodapp. He carefully reaches inside the T2 and takes out roses, chrysanthemums and floral arrangements – and his good-humoured sunflowers. He then hangs a braided heart-shaped floral arrangement on the left sliding door. Is this an expression of his love for his Bulli, perhaps – a decoration that's not even for sale?

A little chat at the weekly market

Finally, it's become a little warmer. Hodapp is slowly warming up as well. Between bits of conversation with his customers and the spry merchant at the next stand, who is still selling bee honey despite being in his mid-80s (the fountain at the market square apparently is indeed a fountain of youth), Hodapp finally tells us more about his Bulli. For example, about the first time he filled it up with petrol: "55 litres for 34 marks 70!" We all groan. Seat belts were retrofitted later on: as we all know, back in 1968 everything was more free-spirited and nobody wanted to be restricted. "I had to replace the engine after 81,000 kilometres – that was a complicated job", Hodapp recalls. "With its second engine, the van now has 194,500 kilometres on the clock." The top speed of 140 km/h was a top value back in the day. We take a look at the cabin: the Bulli is registered as a lorry, and a peek at the cockpit

New joy on every trip. The smile of flower fan Hans Hodapp looks to be nearly as wide as the continuous bench seat

PHOTOS MARKUS BOLSINGER **TEXT** BASTIAN FUHRMANN

The Hodapp garden business is blooming, and one thing stays: the T2

confirms that it is a workhorse. Everything here is original but somehow also looks used, so it wouldn't do as a film prop. Still, the 50-hp engine reliably gets the job done every day, even in winter.

The T2 is used for both flower deliveries and family holidays. "We've got three boys and we often take trips with them and our sledges to the mountains", says Hodapp. "The optional engine-independent heater is a big help here. We also once had a couple of adventures on the way to Laternz in Switzerland. For example, one of the mountain passes was so narrow that when two cars approached from opposite directions, the drivers somehow had to decide who would back up to let the other through."

You can bet the Hodapps were the ones who backed up – they trust their Bulli like it was a member of the family. Hodapp now reveals to us that he has a second Bulli that's also in its original condition and is also used every day. "My wife wanted a Bulli as a wedding present, so I got her one", Hodapp explains. "She drives it to the other market to sell her flowers there." This Bulli is a 1971 T2. "Wait till I'm finished here and then I'll show it to you."

Sliding doors on both sides makes loading and unloading easier. It takes just a few minutes for Hodapp to set up or take down his market stand

Cheered on by the sunflowers

It doesn't take long until we're on our way with Hodapp and his Bulli with Lake Constance dropping back behind us. We come to a seemingly endless field of sunflowers. The green Bulli races past the field, cheered on by the sunflowers and arrives at Hodapp's house.

We walk back behind the plant nursery and pass an old Kramer KL 12 tractor from 1956 (made in Überlingen) on our way to the garage that houses the other T2. The 1971 model is painted lime-green. Dried field flowers hang from the garage ceiling, and a spider has made itself a home across the wings of a few historical model aeroplanes, including a seaplane with a Wankel engine. Apparently, Hodapp uses his T2 to transport aircraft as well! He actually once competed in the German championship for model glider tugging in 1978 with a Graupner Caravelle. He didn't win, though. "It's not the winning but the taking part that counts", he says. Hodapp's wife, Angelika, is now unloading her T2 and is in a great mood. "People didn't buy a whole lot of flowers", she says, "but that Bulli is just so much fun! It's hard to stay in a bad mood with it."

The "flower family" also stays calm whenever one of their Bullis has a technical problem. "With a Bulli especially, fixing things is never a problem", says Hodapp. In any case, the family has no plans to buy a new van. Hodapp's sons also love the iconic automobile. Ingo, who will someday take over the family business, is totally immersed in the T2 theme.

The Hodapp flower business continues to bloom, and it has one constant – the T2. "The two Bullis are simply in my heart" says Hodapp. "Now it's time to get them ready for the next trip to the market." You can be sure the heart-shaped flower arrangement will be hanging on the door once again. Without a price tag.

VOLKSWAGEN
ÜB
HT 46

Hodapp's garage fresco: Elm green between linden green, rose red and cornflower blue. In between a couple of shelved model airplanes from the German championship

VACATION FLIRTATION

New Zealand. A long way from home. No obligations. Just this butterfly feeling in the stomach. In brief, intense fashion, like a summer storm, a 1966 Bulli captured the hearts of two friends from Berlin

Apartment with an ocean view. In the Volkswagen camper, time is relative and a schedule becomes a senseless relic of the everyday world. Enjoying and leaving yourself open to new destinations and adventures – that is the Bulli principle

Green zone, for miles: At some point the Bulli stopped counting how many times it had driven around New Zealand. "At the start of our tour across the southern island, the odometer reading was 99,332 – and after two weeks and around 2,000 kilometres, it didn't move another millimetre."

Type
T1
Model year
1966
Built in
Germany
Power (kW/hp)
approx. 44/60
Engine (displacement)
1.6 l boxer
Colour
Titian red, beige grey
Special features
Temperamental, especially the brakes, named Bertie

Name
Catrin Barnsteiner
Year of birth
1975
Place of birth
Waiblingen, Germany
Place of residence
Berlin, Germany & Oxford, United Kingdom
Profession
Author
Passions
Freshly sharpened pencils, fresh note pads and very special: freshly made plans

If it had been down to me, I'd have simply hired just an ordinary car. Julia, on the other hand, the photographer, insisted on a Volkswagen Bulli, model year 1966, as it had "more soul". In such matters I'm admittedly more heartless: the only important thing to me about our car is that it gets us to all the sights that I have highlighted in my four travel guides – along with a strict timetable.

We pick up the Bulli in Christchurch. Julia takes a seat behind the wheel first, and off we go northwards, right along the coast. Playing out before our windscreen is a sunset of almost criminal beauty – as if the evening sky was on drugs. We are totally enraptured, and the fact that the Bulli brakes don't always work reliably and the petrol gauge, fourth gear and main beam don't work at all – all that somehow fades into insignificance like the setting sun.

Then Julia turns sharply around a corner – and the ignition key falls out of the lock and clatters around the footwell. But the Bulli keeps on going – even with no ignition key. We look at each other, mouths agape, and almost end up in the ditch. If it hadn't been so already, it is clear to us now that whatever it is we've hired, it certainly isn't just an ordinary car!

A few evenings later, while lying on the rear bench seat folded down as a bed, I'm brooding over our schedule: we're too slow! It would be unfair to put the blame on the Bulli alone. We are simply stopping too often. But how could we ever forgive ourselves if we were to simply drive past the most crystal clear mountain lake or most ancient tree or most rugged beach? We have to stop, even if it's the fourth most crystal clear mountain lake or most ancient tree or most rugged beach of the day. Many of the most beautiful bays don't even have a name: there are simply too many of them here.

But the four travel guides with the unvisited tourist sights sit heavier on my lap every day, like admonishments. And eventually even Julia, clearly the more laid-back of us, says: "We need to get a move on."

A little later, as if the Bulli wanted to teach us a lesson, it happens: the clutch cable breaks and we are stranded at a garage in Greymouth, a tiny town on the west coast. Greymouth was most definitely not on my list of tourist sights to visit. It takes days until the replacement cable arrives and we are finally ready to go again, and then I can't find my timetable with all of the stage-by-stage destinations. But who cares? The sun is shining, we start up the engine, open a packet of Kiwi chocolate – and follow the road, like a trusted companion.

A little less than two weeks later in Berlin, in the taxi on the way home from the airport, the taxi driver wants to know all about New Zealand. And whether we'd hired a car there. – "A car? No", I say. "We had something much better – a Bulli!" ❤

Many beaches allow camping buses to park. Then an evening show plays itself out in front of the windscreen – a New Zealand sunset. Clearly too exciting to sleep while it is happening

Has been known to chase down seasoned race cars. Roars like a small fighter squadron. Can hold four people. And it seems more and more peculiar the more you look at it …

RACE-T

No ordinary old-timer: what at first glance appears to be a normal, laid-back Volkswagen Bulli from the 1960s is actually a high-powered racing vehicle with a more than 500-hp biturbo engine taken from the Porsche 993

RECARO
OMP
Fred

Type
T1
Model year
1962 (body)
Built in
Germany & Switzerland
Power (kW/hp)
approx. 397/540
Engine (displacement)
3.6 l Porsche boxer
Colour
White, grey
Special features
Porsche 993 turbo technology with all-wheel drive

Name
Fred Bernhard
Year of birth
1965
Place of birth
Lützelflüh, Switzerland
Place of residence
Utzenstorf, Switzerland
Profession
Automotive expert
Passions
Doing surprising things

It must have been a mirage. Or maybe even an hallucination, as far as I'm concerned. Something like that. In any case, it couldn't have been a Bulli. Alright, it's true that many people like to give their vintage vehicles a sportier look by fitting them with wide tyres and radically lowering their ride height – but this T1 roars like a small fighter squadron and has also been known to chase down seasoned race cars. Its proportions are somewhat peculiar as well. So what exactly is going on here? And why is this cheeky Bulli also sticking its red tongue out at us?

Fred Bernhard is certainly having fun, as our reaction is exactly what he was hoping for: amazement and an incredulous shaking of the head, quickly followed by feelings of fascination and enthusiasm. Indeed, Bernhard's T1 bus always captures the hearts of everyone who sees it. When his "Race Taxi" begins to accelerate with a fierce growl, the last thing you think of is a delivery van with a modest engine. Bernhard's bus sprints from 0–100 km/h in just over five seconds, reaches a top speed of 250 km/h on the racetrack and has braking points that will make the hearts of even the bravest passengers flutter. The four-seat vehicle only looks like a T1, while indeed the technology concealed beneath the lovingly reconstructed exterior skin of the bus is more or less that of a Porsche 993 Turbo.

"I wanted to build something that not everyone has", says Bernhard, a trained coachbuilder from Utzenstorf in the Swiss Canton of Bern. "I wanted to create something sophisticated, something that would be a real challenge. We actually got the idea for the Bulli Porsche one night while drinking beer."

The project really got going in 2002, when Bernhard stumbled upon a 1962 split-window T1 model. However, except for its registration document and the body, the bus was useless. Bernhard therefore got hold of a chassis from a 1985 T3 bus, and – as luck would have it – soon afterwards he came upon a newspaper ad for a 993 Turbo that had been in an accident. The car's technical equipment was in fine shape, so Bernhard bought the car that same day.

A four-way split on the path to a new life

He now had all the ingredients he needed. Naturally, the rusty T1 body wasn't such a good fit for the larger T3 platform – so Bernhard grabbed an angle grinder and cut the body up into four sections, which he later artfully reassembled into a new and larger whole. "I wanted it to still look like a VW bus", he explains. "This wasn't simply a question of making adjustments; there was nothing but air between the components – 21 centimetres crosswise and ten centimetres lengthwise." This posed a huge problem, particularly in the area around the double doors on the side. The

Race bus driver: Fred Bernhard turns very serious as soon as he hits the circuit. His FB1 meets the highest standards when it comes to safety. A total of 42 metres of steel piping serve as a sturdy rollover cage

roof, which required 140 hours of labour, is now made of carbon and is over 30 kg lighter than the original component.

Bernhard's next job was to install the six-cylinder biturbo engine – which he had boosted to 540 hp – in the back of the bus. He replaced the engine's original transmission with a six-speed manual transmission with a cable shifter from the Porsche 996 GT3, which also includes a sintered-metal race clutch and a single-mass flywheel. An auxiliary oil cooler, picturesquely arranged between the rear seats, ensures a comfortable operating temperature for the engine. Bernhard installed the intercooler above the engine in the interior. The intercooler gets the airflow it needs from the side windows in the rear. The air is supplied by hoses that are so large that children could use them as tube slides.

The suspension posed yet another challenge for the "Bulli Builder": "All of the hinge points for the wheel suspensions had to be altered on the platform." A double wishbone design was used at the front, a semi-trailing arm layout for the rear. The suspension turrets were raised 70 millimetres to enable lowering of the ride height by the same amount. The spring-bearing threaded dampers were custom made by Bilstein. All in all, the bus's running gear is as adjustable as that of a race car.

Then, Bernhard turned his attention to the interior. The first thing he did was to take 42 metres of steel tubing and construct a rugged safety cell that would also serve as a load-bearing structure for the body and chassis – after all, Bernhard's plan was to use his souped-up T1 as a race taxi. So, it is not surprising that Bernhard and his passengers – up to three in number – all sit tight in bucket seats from Recaro, each equipped with four-point safety belts. The dashboard and pedals were taken from the wrecked Porsche.

All parts were made as light as possible. Bernhard even drilled holes in the ignition key. He did the same with the passenger foot rests. Bernhard refers to these as "steel panels of fear" because the people he takes along on his drives press their feet down on them in terror whenever they think the 1.5-tonne bus might miss the brake point for an upcoming curve. He always gets it correct, though, Bernhard says with a smile. He then turns on the bus's engine.

The acoustics are irritating. No metallic Beetle clanking, no rattling from the rear - but instead pure 911 sound, unfiltered and rather loud. Pure race car feeling, which applies to the interior temperatures as well

The noise it makes is perplexing at first. There's none of the metallic rattling of a rear-mounted Beetle engine – just the pure roar of a Porsche 911. It's a full-bodied, heavy bass and very loud sound, as there is virtually no separation between the engine compartment and the occupant cell. This also explains the interior temperature. When the bus really gets going, the temperature inside can easily reach 70 degrees Celsius, giving the occupants a true race car feeling.

And that feeling persists during the drive. The passengers don't know what has hit them as the T1 bus with more than 540 hp darts forward. Acceleration from 0 to 100km/h in 5.0 seconds is impressive enough in a sports car – in a Bulli it seems surreal, despite the relatively low seating position. Curves are an extraordinary experience as well, as the old Bulli may look tall, but indeed has a very low centre of gravity. This, in combination with the extra-wide 18-inch slick tyres, enables incredible cornering speeds, while huge disc brakes from the 993 Turbo ensure brutally effective deceleration before entering the next curve. What was that about the foot rests again?

Bernhard spent 3,000 hours working on his one-of-a-kind bus. It took him six years to finish the "FB1", which stands for Fred Bernhard 1, of course. The bus isn't used all that often. Its creator only takes courageous passengers for a spin at a racetrack two or three times a season, although demand is higher than that. That makes each "excursion" a special occasion, and the "mirage" remains what it is: an apparition as rare as it is impressive, one that sticks out its tongue (the red towing hook on the front bumper) at everyone before it rockets past them. ❤

Power plant, turbocharged: in its standard configuration, the biturbo engine delivers 408-hp output from 3.6 litres displacement. In the FB1, power was boosted to over 540 hp, which lets the bus sprint from 0 to 100 km/h in just around five seconds

The upright citizen as a brawny road warrior: Fred Bernhard wanted his T1 bus to be able to hold its own on the racetrack – even against a Porsche 964 Carrera – and it seems that's exactly what he has accomplished

New Jersey
RZV·48A
Garden State

EARL AND HIS PRINCE CHARMING

In the fairy tale, the prince frees Sleeping Beauty with a sword. Ian Brady used a lawnmower. His Sleeping Beauty lived just down the road – and was named Earl

When he was two years old, Ian Brady moved into a colonial-style house with his family. From a grown-up's point of view, there was nothing castle-like about it, but it still gave the impression of being old and a little mysterious – especially so for a small boy. Perhaps it's the house's fault that Ian has always been interested in old things. Ian became an explorer. And Ridgewood, a wealthy and modestly sized community not far from New York could have been created just for him. Ian grew up, wandered farther and farther afield and one day arrived at this bizarre house, just down the road. That's where it was, under a tree, and it gave the impression of being deserted and sad: a yellow VW Westfalia camper van from 1977. Ian fell for it instantly. He would dream of family holidays in the camper bus rather than in a tent. And at the tender age of 12, he decided to persuade his mother to buy the old Volkswagen.

Ridgewood, 28 miles north of New York City. Parked under a tree, just as in those times, but in far lovelier condition: the 1977 T2 Westfalia Camper of Ian Brady and his mother Lesley Fowks

His mother, Lesley Fowks, was not averse to the idea. She made a few enquiries and found out that the owner James Rose had died almost a decade ago. Actually in the same year that the family had moved to Ridgewood. The VW had been standing under the tree ever since, in exactly the same spot where Rose had parked it in the late 1980s when he gave up driving.

"I wouldn't let the bus out of my sight", said Ian, who now installs home entertainment systems. "There was something special about this vehicle that had been standing there for such a long time. I knew that my mother was preoccupied with it, too." Sometimes mothers and sons are very much alike.

Driving across the USA in a camper - from lecture to lecture

Lesley managed to track down Dean Cardasis, a university professor who had helped turn Rose's house into the Rose Center for Landscape Architectural Research and Design. Rose, a pioneer in the field of landscape architecture, had built the house in 1953. The design was an expression of his passion for harmony between build ngs and their surroundings. In keeping with the style of the times, he had built a roof garden on it in 1969 and added a meditation hall as an extension, a Japanese-style zendo. Cardasis and others made the Rose Center more and more popular, the house was renovated and there were guided tours for the public.

In the 1980s, Rose travelled through the USA in his yellow camper bus, giving lectures. He was by no means a hippy, but he still spent a lot of time on the road in his Bulli, which he had bought new in 1977. In that same year, he drove south to Mexicc. The official "Turista" sticker is still

affixed to one of the windows. On more than one occasion, Rose gave lectures at the University of Massachusetts in Amherst, where his friendship with Professor Cardasis developed.

James Rose liked to travel and was very thrifty", Cardasis recalls. "He didn't buy the bus because he liked camping, but because it was so practical for his trips." After Rose died, the house deteriorated visibly. "It was falling apart", Lesley Fowks remembers. "People called it the tree house, because it was built around the trees." The camper van looked even worse. The wheels were rusty, the tyres flat. Spots of rust were appearing everywhere on the once brilliant Chrome yellow bodywork, plants were sprouting out of the rusty overgrown gutters. The engine didn't work.

As a teacher and designer, Ian's mother found the whole story intriguing, but Ian could think of only one thing: the T2b. In actual fact, Lesley Fowks had initially been toying with the idea of buying the bus. Yet she was relieved when Cardasis said it was not on offer after all. "I said to my son: it's a wreck. Apart from that, the man doesn't want to sell it. And you're just 12 years old", Lesley recalls.

Ian continued to negotiate on his own

She then regarded the matter as closed and didn't know that Ian was himself bargaining with Cardasis for the van. His mother was shocked when one day he proudly told her that he would be getting the VW "for nothing". Now aged 13, Ian explained that he had made a deal with Cardasis. He

Earl wears Chrome yellow.
For the former fashion designer and current fashion and interior decoration teacher Lesley, it is pure satisfaction

Type
T2
Model year
1977
Built in
Germany
Power (kW/hp)
51/70
Engine (displacement)
2.0 l boxer
Colour
Chrome yellow
Special features
Two-owner bus; the first owner is the famed landscape architect James in the USA, and the name of the bus was: Earl

Name
Lesley Fowks and Ian Brady
Year of birth
1953 & 1989
Place of birth
Niscayuna & New York, USA
Place of residence
Ridgewood, USA
Profession
Design teacher & entertainment and computer technician
Passions
Camping, beach, cars, computers

would earn the camper bus by working on Rose's land.

"Ian made a great impression on me with his passion for the VW", said Cardasis. "I immediately had the feeling that we should find a way for him to get the bus."

Fowks listened to the details of the transaction and initially nodded her approval but remained suspicious. Every day after school, her son was to spend an hour raking leaves, cleaning fountains and mowing the lawn. At the end of the school year, the camper bus was to be his. Every day, Ian did his work. Afterwards, he didn't go home, but set to work on the camper bus. "He spent hours with cleaning fluids, mould removers and steel wool for the rust spots", his mother recalls. "Everything was filthy, the checked pattern on the seats was barely visible." Ian's determination paid off, and the VW became his. He called his van Earl. There was no special reason for this name. Ian simply liked it. The young VW specialist worked away for hours until Earl was fit for the road. When the bus was finally up and running, Ian, his mother and his younger sister would go off on camping trips. "Compared to a tent, the bus is pure luxury", exclaimed Lesley Fowks, who sees the VW as a family treasure. A treasure that has cost her 9,000 dollars over the years, because Lesley supported her son's efforts in various ways, such as by buying spare parts.

Always still a holiday vehicle – what else could it be?

Ian and his sister are now in their twenties and only use Earl very rarely. Quite unlike their mother, who enjoys summer holidays with her sister Lauren in the VW. By the way, Ian is now restoring a Jeep, which is another vehicle with a history. It used to belong to his grandfather.

Camping-yellow-green. Mother and son share the T2 in traditional fashion. The two are the second owners of the Bulli. The first was a landscape architect. Now Ian and Lesley beautify the landscape around New York with the Type 2

THE SMILE

There aren't many people who know more about the second generation bus than Alexander Prinz – for him the T2 is the number one

It goes without saying that taste is a personal matter – and without a doubt the question of whether the T1 is the best Bulli of all times belongs to this category. Alexander Prinz' favourite bus is the second generation. For him, it is not the original design that is the most perfect, rather it is the successor. He has been passionate about the T2 since 1997 – before then he didn't have anything to do with VW buses at all.

"I've been collecting model cars from Wiking ever since I was a child. But at first I wasn't fixated on any particular make or model", is how the 42-year old sums it up. Until one particular summer evening: "I was still a student, and on my way to a beer garden I saw – from the corner of my eye – a friendly face looking at me". This image flashed into sight from between the high buildings for just a brief moment. It was a T2a, model year 1969. Then the young student discovered a piece of paper inside the Transporter that said "For sale". "I phoned the number right away, arranged a test drive, deliberately overlooked all of the rust patches – and I bought the T2."

Without having any clue about the vehicle. To educate himself, Prinz looked for information and came across the T2 Interest Group. "I got lots of help there, especially in restoring the Bulli." But once you've had every screw and bolt in your hands once, more and more questions arise. For example, how to decipher the extra equipment codes, referred to as "M-codes" that are on the little aluminium badges. When Alexander Prinz, who was born in Cologne, startec working for Volkswagen in the year 2000, new opportunities arose. "I simply asked the right colleagues and I got more and more answers."

This spawned the idea to write a book together with a friend, Vincent Molenaar. The book, VW Transporter and Microbus Specification Guide 1967-1979, written in English focuses primarily on the engineering and less on general facts about the second generation Bulli. However, Alexander Prinz now knows these facts down to the last detail.

As the vice chairman of the T2 IG he is pleased to pass his knowledge on to other bus fans. Including to the youngest of them all. "Although my son Julius is only three years old, he already knows a lot." It's obvious that Prinz junior has been captivated by the friendly face of the T2. And that's no surprise.

Type T2
Model year 1969
Engine 1.6 l boxer, 35 kW/47 hp
Special features Two sliding doors and centre bench seat folds down on both sides

Name Alexander Prinz
Year of birth 1972
Passions T2, family

His great love is family. To Alexander Prinz, family includes – along with his wife Simone, son Julius and daughter yet to be born – his 1969 Clipper N with its two sliding doors. Without a doubt

PHOTOS THORSTEN DOERK **TEXT** GEORG OTTO

THE SOUNDS OF SILENCE

Nature is a single soundtrack. You've just got to listen. Like Gordon Hempton does. He is a soundtracker. A man who is always going out on expeditions to capture the sounds of nature. His mobile sound studio? A VW camper van. What else!

In the surf, Gordon stands and records the sounds of water breaking on the shore in foamy spray

Sound studio and beach house. Gordon's T1 is equipped with everything he needs as a travelling soundtracker. In the winter, a stove chimney extends from the roof, while inside it hums and Gordon goes bodysurfing

What is that man doing on the edge of the ocean? Fishing? Not in that spray, surely! And yet fishing he is. For sounds – the sounds of nature. Nature sings, says Gordon Walker Hempton, a dyed-in-the-wool "Northwesterner". Although he was born 62 years ago in sunny California, he is now a resident of Washington state, up in the upper left-hand corner of the USA. Actually, resident might not be the fitting word for him, as Gordon Hempton, known as Hempy, is continually on the road in search of sounds. "Soundtracker" is what it says on his business card. It is also the name of the movie, to which he has devoted the subject of his life's work, alongside a book entitled "One Square Inch of Silence". For him, therefore, it's all about silence and sounds, nature's score.

What do you need to be a soundtracker? Good microphones. And a feel for the sounds that exist all around us, if only we would listen: the sweeping of the sea, of the wind and the leaves; the cracking and clicking of foliage and rocks; the twittering of the birds; and the background hiss coming from somewhere or other. And the silence in all of that as well, the quiet that the sounds interrupt. Music isn't just made up of sounds; the pauses in between them are no less important.

It is clearly impossible to experience such open-air concerts in the city. They are to be found out in the country, where there are no cars or machines drowning everything out. Out where nature rules is where Hempy goes on his sound expeditions, with ingenious recording equipment, a complete sound studio and his T1 – a VW camper van that is not much younger than he is: the pastel green Bulli was built in 1964.

If you want to sell it – then I'll buy it!

When Hempy first saw it, somewhere in a sailors district of Seattle, the vehicle was already 25 years old. He had been looking for just such a Bulli: top condition, flawless paintwork, chrome-plated wheel hubs, as only a VW can have – and a retrofitted plexiglass roof with curved panes, almost like the Samba buses used to have. Only, on Hempy's camper van the skylights are bigger.

Gordon Hempton placed one of his soundtracker business cards behind the windscreen wipers: "If you want to sell it, I'm buying!" He did that more than once, but never received a reply. And then, five years later, this Bulli he so wanted to own suddenly crossed his path. And in the back window was a sign: "For sale"! Dan, the owner, needed persuading. He was his

PHOTOS CHRISTOPHER LAMARCA, KARI WILTON **TEXT** GORDON HEMPTON

Listening to the forest grow. What looks like taking a nap is in reality hard work. Gordon is adjusting his sense of hearing to perceive what is moving in the forest

Type
T1
Model year
1964
Built in
Germany
Power (kW/hp)
30/42
Engine (displacement)
1.5 l boxer
Colour
Light green
Special features
Mobile sound laboratory & camping retrofit with wood stove

Name
Gordon Walker Hempton
Year of birth
1953
Place of birth
Fort Ord, USA
Place of residence
Indianola, USA
Profession
Environmental acoustics researcher
Passions
Body surfing, recording natural sounds, the bus

Now there is even a movie about the sound researcher. A sensitive report documenting a person who listens precisely- and experiences nature so very differently than most people

Bulli's best friend and wanted to be absolutely sure that the new owner would at least be its second-best friend. Gordon's assurances that he had already owned two VW camper vans did not impress Dan at all. And so, Gordon took a deep breath and declared his personal Bulli creed: "Check the oil level before you turn the ignition key and check the valves every 1,000 kilometres, always with the engine cold!" And underlined his speech with expressive fingers: "The feeler gauge has to slide like a knife through soft butter." Two days later, he was the truly proud owner of the camper van for which he would have paid any price. In fact, he had to shell out 2,500 dollars for the bus, toolbox and logbook included.

In parting, Dan gave him three rules for this Bulli icon. Firstly: forget about hurrying anywhere, as 80 km/h is pretty much the limit. Secondly: never assume that you will get to where you're going. So enjoy the journey all the more. And thirdly: you will need a whole host of friends to help you to search for parts.

Luxury today means being removed from the environment

Searching for parts would not be a problem. Gordon Hempton has travelled all over the USA with his camper, from Washington state, to Washington, the nation's capital. Nature sings its polyphonic song wherever we let it. Nowadays (none of us are getting any younger, the Bulli included), Hempy prefers shorter trips. Always on board are his recording equipment, with an array of highly sensitive microphones, and a sound studio, in which he can process the day's sounds locally and send them out via iTunes into the wide world of the web.

There you can hear the roar of mighty freight trains, rolling thunder, pattering rain and crunching snow. Chirping crickets and hooting owls. As well as quiet, the "sound of silence". Gordon Hempton literally hears the grass grow. His "Soundtracker" film, which features the Bulli veteran in fitting fashion as well, was produced by a company called "fou films" - crazy films! And indeed, many people also regard Gordon Hempton as a bit crazy. At times, he reinforces this impression of himself, for instance by starting a campsite belching competition with his family – wife Julie, daughter Abby and son Oogie – within earshot of their distraught neighbours. A competition that he, of course, regularly wins! Or whenever he jumps into the waters of the Pacific to go body surfing and lets himself get carried along by the waves. Wearing just a neoprene suit and with small boards strapped to his hands. Gordon experiences nature in his own way, is a part of it, trusts it and listens to it. Perhaps the best way to approach nature is in a purist T1 that has a small cooker inside, so you can warm yourself up. No excessive luxury that sets you apart from the environment. No unduly heavy footprint pressing the tyres into the earth. Simply a Bulli – what else!

VOLKSWAGEN – FAMILY CARS

SUM – NOT PART OF THE WHOLE

It never imposes, it's a guardian angel, a best friend, a place of retreat, a getaway car, the finest backdrop, the most mysterious playground, a pal, a brother, sister, a dandelion clock. Capturing the aura of the Bulli is every bit as easy as catching a handful of soap bubbles. The bus from Hannover is a legend, without ever trying. It is honest, authentic, original, unaffected, politically incorrect, yet always fair. Falling stars are also there simply to be enjoyed, to wish on everything worth having, and those who sing of putting them in their pocket for a rainy day are just doing so for the sake of the finest poetry. In the years to come, one thing will remain in the memories of these budding Bulli fans, these little people with their big eyes and infinite chains of thought, and that is a basic sense of trust in one enduring constant: the Bulli.

PHOTO-ESSAY FRANK KAYSER (FOR LILLY, ELLA, MATTEO UND KATARINA)

Can you see me? I'm exploring a continent of surprises. Come on, jump in – there's a very special smell here

All is well. Lost in dreams. You can almost feel the world slowly growing warmer behind the curtain – safe and secure, always

BN 724 FX

First one up. Everything covered by a light, moist film, toes inside tiny socks are getting damp in sandals – it doesn't matter, I'll just keep leafing through my book until the grown-ups are finally awake. I wonder what's for breakfast

All-inclusive corral. Bullis of all generations unite: we are the Volkswagens!

Anyone who can guess what a tremendous number of star fleets have conquered space behind this brow will also know what a precious freight it carries – fortunately in the best transport in the universe

FACTS & FIGURES

Model	T1	T2	T3
Year			
1949	8		
1950	8,059		
1951	11,979		
1952	21,496		
1953	28,177		
1954	38,493		
1955	44,255		
1956	50,128		
1957	81,396		
1958	92,777		
1959	112,467		
1960	128,505		
1961	141,662		
1962	157,845		
1963	168,241		
1964	179,916		
1965	170,215		
1966	169,648		
1967	134,378		
1968		215,267	
1969		230,162	
1970		241,865	
1971		233,563	
1972		233,939	
1973		216,965	
1974		137,846	
1975		129,118	
1976		140,219	
1977		150,232	
1978		141,307	
1979		72,435	49,361
1980			151,659
Total	1,740,245	2,142,918	

Model	T3	T4	T5
Year			
1981	124,786		
1982	113,195		
1983	114,565		
1984	114,274		
1985	106,366		
1986	114,394		
1987	107,785		
1988	101,556		
1989	100,422		
1990	79,306	23,200	
1991		137,682	
1992		167,798	
1993		129,766	
1994		144,199	
1995		141,709	
1996		142,503	
1997		155,117	
1998		154,980	
1999		148,883	
2000		162,709	
2001		151,970	
2002		137,762	219
2003		53,285	92,171
2004			136,101
2005			164,132
2006			183,790
2007			186,867
2008			185,313
2009			107,323
2010			250,480
2011			162,825
2012			158,948
2013			150,048
2014			164,655
Total	1,277,669	1,842,563	1,834,881

T1
1,740,245

T2
2,142,918

T3
1,277,669

T4
1,842,563

T5
1,834,881

BULLI VITA

1950	Production of Volkswagen Type 2 T1 begins on 1 March.
1955	Bulli gets better ventilation and therefore gets its characteristic roof projection
1958	Production launch of the VW pickup with double cab
1960	Starting 1 June, Bulli has blinkers instead of trafficators
1963	VW bus gets a wider rear lid. Unfortunately, this eliminates the corner windows of the Samba version
1967	Model change; T2 bus replaces the Bulli with split windscreen
1972	New reinforced frame, new box bumpers and other refinements make the VW bus safer
1978	First test drives with an all-wheel drive model, the forerunner of the Syncro model that was later produced
1979	Introduction of the new T3 bus, which is more angular and larger than the previous model
1981	First diesel engine in the VW bus; it has a power output of 50 hp
1982	First water-cooled petrol engine in the VW bus; transition away from air cooling
1985	All-wheel drive Syncro makes its debut in the VW bus
1988	"Magnum" special edition model arrives on the market – a very well-equipped version
1990	The T4 arrives! Now with front-mounted engine and front-wheel drive
1996	Comprehensive facelift of the T4; the VR6 engine is now also available in the bus
2003	T5 makes its debut as fifth generation of the Volkswagen Type 2
2009	Facelift of the T5
2015	The T6 is presented – the sixth generation of a global success

BULLI FIGURES

5,850	Deutschmarks – Price of the Volkswagen Type 2 in 1950
9	Maximum number of seats in a VW bus
760 KG	Payload of the first Bulli
1956	Opening of the Volkswagen transporter plant in Hannover
100 DM	VW stock price when it was first issued in 1960
L53	Colour number for sealing wax red, one of the most beautiful colour tones of the VW T1
1,455,034	Number of T1 buses built in Hannover from 1956 to 1967
7	Number of switches and controls on the standard dashboard of a T2 bus in 1967
25 HP	Power output of the first VW bus engine in 1950
M113	Ordering code for Safari windows, i.e. the folding windscreens of the T1
2	Number of VW buses exported to the USA in first production year 1950
112 HP	Power of the most powerful production VW T3 bus (2.1-litre water-cooled boxer)
50° 30'	Mounting angle of the first diesel engine in the VW T3
6	Number of Bulli generations
241	Precise model designation for the Samba bus as a nine-seater
1969	Woodstock
235 HP	So far the highest engine power available in a production VW bus: the T5 3.2 V6
65	Number of years the VW Bulli has been built – longer than any other comparable commercial vehicle

#BULLILOVE

20 DECLARATIONS OF LOVE IN 140 CHARACTERS THAT TOUCH THE HEART

T2, 1995 / **Inácio Rozeira** / The man from Portugal loves this 9-seat Brazilian built bus. On one trip to Bolivia, he climbed to an elevation of 4,600 metres with the Bulli

T1, 1967 / **Steven van Schoor** / The Belgian likes to live up high. His jacked up Bulli ploughs over many a field this way. Inside, his family finds plenty of space for travel

T2, 1979 / **Eric Sell** / The man from Namibia heads on a regular basis out onto the plateau to rescue abandoned Bullis. His fleet now includes ten desert finds – from T1 to T2

T1, 2008 / **Kirk Strawn** / created "Walter The Bus". The world's largest Bulli is a regular fixture at the annual Burning Man Festival in the Nevada desert

T2, 1969 / **Mariana Abaroa** / The woman from Mexico got "Shaggy" from her father. Shaggy is often seen waiting in front of the small health food store in the Santa Monica Mountains where Maria earns money for petrol

T1, 1967 / **Johann Lafer** / is a master chef and has already earned accolades from the Gault-Millau and the Michelin Guide

T4, 1991 / **Rolf Pähler** / from Cologne loves motorcycle riding – and Bulli driving. Ideally in the same outfit

T1, 1964 / **Bernd Berger** / This T1 did a gig by unceremoniously riding up to the top of the Austrian Kitzsteinhorn mountain by ski lift. To an elevation of 3,203 metres. Followed by drums and beats

T3, 1984 / **Gino Messinas** / and son enjoy a morning surf in the Pacific near Los Angeles. Practical: their T3 has an outdoor shower

T1, 1968 / **Jason Cooper's** Voodoo Bus was exclusively intended for saving money for a wedding. Under the Welsh sky one often hears: "Yes, I do."

T1, 1963 / **Mario Mutenthaler** / either travels with his family or to his other family. Like to this Bulli meet in Kals am Großglockner, Austria

T1, 1955 / **André Birrer** / This pickup was originally delivered to the USA, but in 2003 it returned and sailed over the Rhine to the welcoming hands of Andrè in Basel, Switzerland

T1, 1957 / **Eight apprentices from Linz** – and one Bulli: an ingenious apprenticeship project at an Austrian official Porsche garage. "We could have had no greater joy than this."

T1, 1964, 1965 & 1967 / **Carsten Schiedrum** / This isn't a press photo; this is real work. The roofing company from Eschwege uses this trio to climb roofs

T2, 1969 / **Lena Schrapper** / The 25 year old from Oldenburg travels alone around the world. Her only companion through Panama, South Africa, Australia and New Zealand: this mint green T2

T1, 1954 / **Ken King** / Canadians might be aston shed when this rare folding roof barn door model folds out its direction indicators for turning

T3 Syncro, 1986 / **Chris Killian** / In 1984, the American journalist set out in his T3 to discover the pop culture of his homeland. This photo was taken alongside highway US 50, known as the "Loneliest Road in America" in the state of Nevada

T1, 1965 / **Rolando Alverado** / A turbo engine with an output power far in excess of 200 hp makes the regular excursions through the Golden State taken by this Californian and his family sweeter

T1, 1965 / **Werner Herzenberg** / During his journey across Swasiland, the South African and VW enthusiast provided sufficient cooling for his favourite car from Hannover

T1, 1963 / **Bob Marley** / Trenchtown Cultural Centre, Jamaica. He lives on! At least his bus, with which he went on tour with the Wailers and took his children to school

This is art: with its significantly honed contours, the T6 is an eye-catcher that looks good in a gallery. Design and utility combine into one unit

BUILT WITH LOVE

Nothing is harder than following in the footsteps of a legend. The T6 solves this problem in an elegant way: a touch more charm outside and completely new inside. An understated bus in the best sense of the term, and one that doesn't simply rely on its predecessor's reputation but continues the journey

As Chairman of the Board of Management of Volkswagen Commercial Vehicles, **Dr Eckhard Scholz** is one of the fathers of the new T6. He sees the recipe for success in the Bulli's variety and versatility

The Generation SIX model in two-tone paint finish is a tribute to the T1 Samba. A special model based on the Multivan Comfortline to coincide with the launch of the new T6

Which Bulli should it be? Only when they reach full size do the two-tone Generation SIX models achieve their full effect. A picture to dwell on

H NT 600

Everything on the inside really is new. The dash panel comes in two versions: one in a luxurious design, the other with more basic equipment and a larger number of open compartments for tradesmen. The seats, shown here with Alcantara, exude a premium-class flair

Red and white and large. Unmistakably Volkswagen Bulli. Functional lines. A design revolution would have been unnecessary, because a well-organised and spacious design fit the bill and are now packaged more attractively. The first encounter with the new T6 is very amiable. As if you were meeting up with an old acquaintance who has been travelling around the world for a few years and is now suddenly standing before you. His facial features may be tanned by the weather, more distinctive, sharper, but he is still the same old likable character that you know.

The new clothes suit him well. The cherry red and the candy white, for example, look great and hark back to the 1960s. That was when sealing wax red and pearl white were popular colours for the Samba bus, the equivalent of today's Multivan and Caravelle. Fans of the classic models will miss the folding roof, but then that has been absent from the Volkswagen Bulli since 1967. Today, the T6 nearly always has an air-conditioning system on board, which is plainly able to do more than the folding roof of yore. After all, time hasn't stood still. More or less immovable seats at the front? Rear benches without any lateral support? Long gone. The T6 is a space wonder as a Transporter and a model of comfort as a Multivan. Now in its sixth generation, it is fully prepared to kindle a new love.

When Volkswagen Commercial Vehicles introduced the front traction unit principle in 1990 and bade farewell to the rear engine, many fans felt a sense of betrayal. Yet a Bulli simply has to be practical in all aspects, and that is why its forebears in Hannover were simply unable to continue along the path once devised by Ben Pon and Heinrich Nordhoff. It was time for new ideas. Today, the crocodile tears from back then have long since dried. The Bulli has reinvented itself. And at the many Volkswagen meets, T4 and T5 models line up next to the classics as a matter of course. The T6 will soon be parking alongside them as well, once the fans have had a chance to explore it for themselves. We are having a look at it a little earlier, and we enter the interior, because here everything is new compared with the

PHOTOS VOLKSWAGEN COMMERCIAL VEHICLES / FOTOLIA.DE **TEXT** THORSTEN ELBRIGMANN

predecessor. Cockpit, door trim and seats – they have all been modernised and made more comfortable. This is particularly so in the Multivan and Caravelle, which differ from the more basically equipped Transporter even in the layout of the cockpit.

Simple craftsman's car or luxury ride - everything is possible

Open storage areas, a narrower centre console and, if required, a dual seat bench are signs of the utility demanded by Ben Pon and remain important up to the present day for the tradesmen's first choice: the T6 as a Transporter. Yet just as the Samba bus exuded luxury during its time, the Multivan now impresses with features such as its high-gloss surfaces and options such as electrically adjustable seats, a tailgate that opens and closes electrically, heated front windscreen and digital connectivity – features one might expect to see in chauffeur-driven limousines. But what is the Multivan if not precisely that? Running mobile phone apps on the multimedia system? No problem. You can upgrade the T6 with every conceivable equipment feature to make the long holiday journey relaxing for everyone. More relaxing for the children, who have something to occupy them, and therefore for the parents too. To ensure that the long journey to the business appointment is not wasted time and that the very latest draft can be emailed into the wide world. A camping version will follow as well of course – that's logical. A VW Bulli can in fact do everything. Yet it conquers hearts above all else. Even after 65 years, the Volkswagen remains true to itself: as a pickup or dual cab, small touring bus or business vehicle, ambulance, fire brigade, cherry picker or camper. Yet that is only half the truth, because for those who experience it day in, day out, it is a saviour, a daily helper, an office, a children's room, a mobile holiday resort. From the modest small diesel to the powerful 204-hp bus, with adaptive chassis, a lowered body or increased ground clearance. With front-wheel or all-wheel drive – drivers can put together precisely the Bulli they want. What the fans around the world do every day when they restore and enhance their old buses is just what Volkswagen does, namely build the best bus. For every single Bulli driver. After all, when all the rationality that can be so elegantly explained in terms of payload, luggage space or versatility is put aside, this is the only way to generate the one emotion which, more than any of the other commercial vehicles in this world, only a VW Bulli can give: love. ❤

WESTFALIA
H VW 774H
Oldtimer
H
VW167H

AS GOOD AS NEW

Life has been tough. But in their old age, they get a second chance – at their birthplace in Hannover. The workshop of Volkswagen Commercial Vehicles Oldtimer brings the shine back to old Bullis

On the hoist, in the hall, in the yard: classic Volkswagen buses are everywhere. T1 alongside T2, T3 alongside T4. There's a smell of engine oil, mechanics are busy adjusting valves and repairing brakes. Like a surreal vision, a bare T1 body passes at some distance on a mobile platform. "We always start by removing body paint and rust for our restoration jobs in an immersion bath," explains Dr Ralf Meyer, Head of the team at Volkswagen Commercial Vehicles Oldtimer in Hannover. It is a gentle method of treatment, which sometimes reveals stamped markings on the panels. They are like a message from a distance past when the gleaming silver-coloured sheets were stacked alongside the production line waiting to be assembled. Covered by paint for decades and now visible again – but just for a short time, because the bodywork artists set to work on the vehicles, removing the traces of previous restoration attempts, cutting away the rusty metal, taking measurements and welding in new panels, which are often fabricated by hand.

There are experts in every field: panel-beaters, who have learnt their trade from the bottom up, engine builders on first-name terms with every bolt, and then there are the chassis specialists, upholsterers, painters and fitters. A museum is one thing that Volkswagen Commercial Vehicles Oldtimer does not have. "Our museum is the road. It's important not to have the Bullis we have here just standing around gathering dust. They appear at classic car events", says Meyer, describing the classic concept of the Bulli workshop at Volkswagen Commercial Vehicles Oldtimer. At trade fairs, the Becks Bier bus and the VW bus with a curried sausage body are still in operation and performing their services without delicate treatment, but they are perfectly maintained to stay fit in a ripe old age.

Apart from the buses for the company's own use, many Bullis in the workshop are being restored for external customers. The experts at Volkswagen Commercial Vehicles Oldtimer begin by consulting the factory archive to determine the original condition. They can then take stock and come up with a precise cost estimate. "The fans appreciate the quality and receive a certificate for the vehicle when the job is finished", Meyer points out. As a result, a restored Bulli from this very special workshop in Hannover is always an original. Forever young.

With great care and great knowledge is how the experts at Volkswagen Commercial Vehicles Oldtimer handle aches and pains, small and large, of the aged boxer buses

PHOTOS VOLKSWAGEN COMMERCIAL VEHICLES **TEXT** THORSTEN ELBRIGMANN

CAPTAIN KIRK AND HIS WOMEN

Loaves leave scars, maintains Jean Viljoen from South Africa. Since his childhood days his imagination has revolved around VW Bullis. From far-off galaxies to Elvis, from hippies to family - it all relates to a Bulli

My first memory of our VW Bulli is still very fresh, even though it is now many years back. I am four years old and going on holiday with my two brothers and my parents. I am standing in the gap between the front seats of the T2, holding tight to the seat backs. That was my place in the Bulli. The bench seats in the back belonged to my older brothers. One seat each. I had the gap. Always at my post. I was Captain Kirk and the Bulli was the Enterprise. Heading to far-off galaxies. I had the long gear lever and the handbrake under the dashboard in view and used to follow the windscreen wipers whenever it rained, standing on tip toes so that I could see through the part of the screen that got wiped clean. My father behind the wheel was the First Officer, who got us safely to our destination. Every destination. Always on board, along with my brothers on the rear bench seats and our treasure chest in the luggage compartment above the engine, kitted out with toys, were a quilt warmed up by the engine and sometimes a musical instrument.

I don't know why, but I also remember very clearly that morning in 1977 when the news of Elvis' death came over the radio. I was five years old and standing at my post, when the announcer said that the King had died in Graceland. That year nothing else happened that I'd be able to remember, but that news item I heard that morning in our VW has stayed with me. I clearly remember the road we were on, I can still smell the air and see the sky.

Summer in South Africa. The 1958 pickup with Safari windows is completely unrestored. A farmer once used it as his Sunday car

Fortunately, children are very resilient

The T2 was our family car. In his job as an engineer my father drove two further Bullis: one with a double cab and one as a flatbed. We lovingly christened the three of them "Loaf", "Half Loaf" and "Quarter Loaf". What has remained with me as my most intense memory from my early years is my thumb going blue for ages, as I liked to hold onto the B-pillar and pulled Loaf's sliding door shut at the same time. Whack! Blue thumb and streams of tears. Two other scars remain to this day: a reminder of falling out of Loaf's sliding door. Fortunately, children are very resilient.

Fast forward 30 years. I am in my mid-thirties and living in South Africa, a part of the world blessed with sunshine, beautiful countryside and an untouched coast. Volkswagen has played an important role in the country's motoring history. Beetles and Bullis were built here from very early on. South Africa was an important VW export and production country. The sturdy cars from Germany were and still are sought after, and they are a common sight on South African roads. People grew up with VW – moreo-

Miz Daizee is the colourful work of an old hippie, and today it lives in a commune with many other buses – and with the Viljoen family

ver, with its friendly face a Volkswagen was one of the family. No wonder that there is a big fan community here. And I am a member. With a great passion for the T1, I am an avid collector, restorer and driver of this vehicle today.

My first project was the restoration of a 1975 Kombi. This model with a split windscreen was still built in Brazil for many years and was imported to South Africa in 1975/76 as a cut-price Bulli. The Kombi was available as a panel van, a flatbed pickup truck or with 15 windows. I drove "Rosie" – my first love, if you will – for many happy years and always with two kayaks on the roof.

After Rosie, I had lots more Bullis, lots of lovers, if you like, for each one was unique and always adorable. But in 2006, I bought one that will remain with us forever. This very special Bulli came into my life in a peculiar way. During a conversation with an old hippie in Cape Town, a friend heard about this colourful 'Woodstock altar'. The owner at that time had the 1960 Bulli from Germany fitted out and painted as a camper. However, when the work was finished he felt it was too good to spoil by driving it. He named the vehicle Daisy, parked it safely in his garage – and forgot it. I bought it months later and overhauled the entire chassis, brakes included. As an expression of my respect for the old lady, I gave her the name Miz Daizee.

My girlfriend (who soon became my wife) and I drove many kilometres in Miz Daizee and explored the side roads of South Africa. We even spent our honeymoon with her. And, naturally, our now four year old son Lucca experienced his first Bulli outing in Miz Daizee. Thanks to the T1, I made life-long friendships with people whom I never would have met without the VW. That's due to our shared love for the vehicle, for this life – and to our wanderlust.

One of these friendships was with Truus van der Mark. She had an alert mind, a big heart and a very unique view of the world, which she had once travelled around in a Bulli. When I met her, she was over 70 years old, but eternally young in heart and mind. In the early sixties, she had set out

FLOWER CHILD

Type	Name
T1	Jan Viljoen
Model year	**Year of birth**
1958	1972
Built in	**Place of birth**
Germany	Colesberg, South Africa
Power (kW/hp)	**Place of residence**
25/34	Paarl, South Africa
Engine (displacement)	**Profession**
1.2 l boxer	Marketing manager
Colour	**Passions**
Sealing wax red, beige grey	Cars and motorcycles of the 1950s and 1960s
Special features	
Has crossed three continents	

with her husband and a 1958 sealing wax red and beige/grey Samba bus, and over the next two years they crossed three continents. We listened to her stories for hours, especially those about her journey through Africa. An album full of newspaper cuttings, faded photos and a 16-millimetre film made her memories more vivid still. Truus has sadly since died. But my family has the good fortune of being permitted to preserve her beloved Samba for coming generations.

Excursion to the sea: Jean's buses are used, but are also lovingly cared for. The right-hand steering models have their folding doors on the left

A few years ago, we discovered a T1 flatbed from 1958. My father remembered a farmer who used to drive his VW as his 'Sunday car'. We found the flatbed in a barn. In the early sixties, there had been a problem with the engine. The farmer had taken it out and put it down on the flatbed. When he died, the Bulli remained that way – for 45 years. For a modest price I became the new owner of a very original, powder blue pick-up from 1958. There was not a lot to do: a new pump membrane for the fuel pump, a general overhaul of the brakes and a thorough cleaning. It was incredible: the Bulli ran like new – but then again, it had not worked on the farm for a single day.

Over 50 Bullis – over 50 love stories

Over the past decade, I've had more than 50 Bullis. I loved each of them for their unique histories, and parting with them was always a sad occasion. However, a few noteworthy examples fortunately remain in my care. In South Africa there are only a handful of Samba buses. We like to drive ours frequently. We live in a wine growing region in the mountains, where 23 windows are ideal for enjoying the view.

Miz Daizee is and remains our camper. Our young children are always excited whenever we set off with her. Miz Daizee is small – but comfortably equipped. And she is slow. That gives us the opportunity to teach our children that half the fun comes from just being on the road. I harbour the secret wish that our wanderlust and love of country roads will take hold of our two children just as much as it has us. Captain Kirk's post on the bridge has now been unoccupied for quite a while. But I think little Lucca is ready to take over this role. With his father as First Officer and driver. ❤

With this Samba from 1958 Truus van der Mark travelled on three continents and then gave it to Jean. She became a good friend

Wiener Riesenrad
RIESENRADPLATZ
PORSCHE
PORSCHE
SALAMU

BULLI BY PORSCHE

"This bus has a fantastic story", says Heinz Swoboda, and his attentive eyes light up. He is a seeker, someone who gets to the bottom of things. And who finds things that no one else sees. This is probably the only way to acquire a VW Bus T3 with Carrera technology

Heinz Swoboda's first classic car was a Pininfarina Coupé, a French car with an Italian design. Old and rare cars have always fascinated him. He turned his passion into a profession, and in his spare time he organised very successful classic car rallies in Austria. There you get to know people, hear stories, rumours of classic cars slumbering in barns and waiting to be discovered. Heinz Swoboda followed up on many of these stories. Hunting down rare cars became a daily occupation. "Sometimes I feel like a truffle pig", Heinz says with a smile. Finding something special, rarities and cars that have long since become extinct has become his trademark.

"I am especially interested in underrated or as yet unrecognised models; I enjoy being ahead of the trend", says the collector who likes to go his own way.

With Fox tyres and a Carrera engine, this Porsche bus parks in front of the big wheel – no seat for a third person on board, but it has six cylinders

This is how it was with his first VW bus B32, which was offered for sale for over three years – unloved and unrecognised by the market. It was a German vehicle, parked in Switzerland, which had already failed to sell at several auctions and was overlooked by key collectors at shows. "I discovered the bus, and I immediately drove to Switzerland. A week later the rare vehicle was in Vienna, and my world was in order."

What does B32 actually mean? The "B" stands for "Bus" and the "32" for a displacement of 3.2 litres. The owner's manual dryly states an incredible fact: "This vehicle is equipped with a Porsche 911 Carrera engine. As a result of the engine power of 170 kW (231 hp) the vehicle accelerates from 0-100 km/h in approx. 9.6 seconds and has a top speed of around 185 km/h. This performance greatly exceeds that of the standard VW model. When driving the vehicle you should consider this and adjust your driving style accordingly."

B32 prototype – it doesn't get any rarer than this

The secret has been revealed: this is a Volkswagen with a Porsche engine and Porsche engineering. A symbiosis that is common today but was very rare back then, which led to some surprises on the motorway. A rare animal, a real character. "There are probably only nine B32s and three prototypes in existence – so a total of twelve buses powered by Porsche. We are talking about a factory-built design and not subsequent tuning. By chance, I found a second B32 here in Austria", Swoboda reports. Tough negotiations followed before the actual star of our story arrived in Vienna. Because it is

PHOTOS BENGT STILLER **TEXT** ARNO MICHAEL HASLINGER

Heinz Swoboda has a special eye for tracking down rare cars – that is what he loves to do!

Type
T3
Model year
1984
Built in
Germany
Power (kW/hp)
170/231
Engine (displacement)
3.2 l Porsche boxer
Colour
White
Special features
Porsche Bus B 32 prototype

Name
Heinz Swoboda
Year of birth
1969
Place of birth
Vienna, Austria
Place of residence
Vienna, Austria
Profession
Self-employed
Passions
Cars that tell a story

one of these three prototypes, and to the present day it is the only known survivor. A VW bus could hardly be any rarer.

Was this vehicle planned as a play-thing for the directors? Or were there specific reasons for developing an "über-bus"? Swoboda found the answer: during test runs in Algeria in 1981, Professor Helmuth Bott, Board Member for Development at that time, initiated the development of a spare vehicle with a 911 engine to serve as a minibus at Le Mans or as a service support vehicle, for example in North Africa. Production of the special model started in the autumn of 1981 – and it was equipped with a 3.0 SC engine featuring K-Jetronic. Two years later, in November 1983, it had already followed test vehicles at high speed for 58,000 km. Back home, the B32 was upgraded to a 3.2 litre Carrera engine.

The powerful Bulli caused a sensation on the roads, and the Americans also showed interest in it. In 1983, production of nine vehicles started with the order number 970/52 and type number 2656. Porsche was the manufacturer, and the 17-character type label started with WPQ ZZZ B 3 ZE S2 0001. By contrast, the internal prototypes had a VW vehicle identification number. They were produced at the Welcherath Porsche workshop at the Nürburgring racetrack. Unfortunately, after a strategic realignment that included further development of the 911, the development centre no longer had any available capacity, and the "B32 project" soon fizzled out.

Our prototype later served as a support vehicle for the Porsche Carrera Cup organisation. Heinz proudly shows the vehicle registration document with the original location code BB-PW 806 and the first registered owner Dr. Ing. h. c. F. Porsche AG. Distinguishing exterior features on the B32 are the side inlets for proper ventilation of the Carrera engine, and the typical Porsche Fuchs wheels measuring 6J x 15 at the front and 7J x 15 at the rear. The drivetrain, brake system and gearbox were borrowed from the Porsche 911 and adapted accordingly.

On safari in the purring character

But how does such a powerhouse drive? Take the driver's seat – a stylish sports seat from the 928 – and put your hands on the leather-covered Porsche steering wheel. You hear the fine whirring of the six-cylinder engine as soon as you turn the key – and the dream of every Bulli enthusiast revs up. The pairing of the usual long shift travel in a Bulli with the elemental power of an air-cooled Porsche engine suit one another like a clumsy elephant and a lithe antelope, yet both drink from the same waterhole, and it is only together that they make a safari a real experience. The original B32 gallops forwards with brute force. From a speed of 120 km/h the wind noise enables the aerodynamic resistance of the cuboid on wheels to be experienced acoustically. One question arises, but it remains unanswered: Heinz is not revealing the vehicle's fuel consumption. After a wild ride, we park the T3 back in the heart of Vienna. We get out at the Big Wheel of the Prater fun fair. A legendary place for a bus with such a fantastic story. And thanks to Heinz Swoboda it's nowhere near finished yet.

PORSCHE
BB PW 806

THE PERFECT WAVE

California is far away, the hippy days are long gone. But freedom and friendship are still alive everywhere. In Japan, a group of surfers are travelling in their Bullis with the Pacific in view

Always with a surf board on board, finding relaxation in the achievement-oriented Japanese society. Outsiders? No, but they do plan small escapes. That is what Kakujo Tomoyuki and his friends like to do

The magnificent seven on the road in the T2, T3 and T4. The main thing is that it be a Volkswagen Bulli, because it is a symbol of freedom for the surfing crew.

Go West! Not to the legendary Bulli land of California, not to LA or San Francisco. Much farther west, all the way across the Pacific to Japan, where west becomes east again. You then travel down the Honshu coast past the surreal beauty of Mount Fuji to Hamamatsu. The busy industrial city, located 250 kilometres southwest of Tokyo, is particularly well-known amongst motorcycle fans. This is the home of Honda, Suzuki and Yamaha. Hamamatsu is also famous among gourmets – for excellent mandarins, eel and fugu. The latter is a fish with deadly poisonous innards and is prepared only by chefs with many years of training.

The bustling of the city contrasts sharply with the peaceful surroundings and sights of natural beauty. Tree-covered mountains are 20 kilometres to the north, the delightful river Tenruyu to the east, Lake Hamana to the west. And to the south there is the Pacific with its sandy beaches between Omaezaki and Irago that are kilometres long. It offers perfect conditions for surfers and Bullis under a friendly and warm sun.

A community of friends lives, works and enjoys life in Hamamatsu with the Bulli as an indispensable source of pleasure. Kakujo Tomoyuki has a role akin to the leader of the gang; he is also a passionate surfer and owner

PHOTOS SHOUTA KIKUCHI, FUMIHIKO ISHII **TEXT** KAZUHIRO NANYO

Type
T3
Model year
1991
Built in
Germany & Austria
Power (kW/hp)
82/112
Engine (displacement)
2.1 l boxer
Colour
Wine red
Special features
The perfect surfing Bulli for any surface

Name
Hisanobu Hayashi
Year of birth
1974
Place of birth
Hamamatsu, Japan
Place of residence
Hamamatsu, Japan
Profession
Hair stylist
Passions
Surfing

of CO.NNECT, a high-end shop for trendy clothing and accessories. A short while ago, he organised a surfing trip and didn't have to waste any time persuading his Bulli buddies. Also joining them is Mako, an old friend from middle school. Satokatsu has been one of his surfing pals for ages, while Hayashi is a friend of a friend. Hidekawa works on the same street where Hayashi runs the boutique GEE. When word of the trip gets around, Kei and Kawashima also pop up. The magnificent seven meet up outside Kakujo's lifestyle boutique and pack their surfing gear into a T2, a Vanagon and a Eurovan. Then it's off on a two-day safari, with a surfing spot near Omaezaki as the destination.

In the evenings, they sat together in their wagon train

Omaezaki is situated at the tip of a peninsula on the mighty Pacific and offers excellent conditions for surfing all year round. The first stop is a place called "Shark". A few years ago, apparently, a number of these feared predators were observed patrolling in the waves. Undeterred, the surfers wax their boards and squeeze into their wetsuits. "I'll go first", says Satokatsu, tucking his Danny Hess board under his arm and then trotting off across the sand. The others follow shortly thereafter. At some point, the lads pack everything back into their VW vans and set off in search of an even better wave.

As the sun sets, the convoy reaches a camping site in Katahama, where they arrange their VWs in a circle – covered wagon style. The Westfalia pop-top roofs are opened. The internationally famous graphic designer Ken Kagami has decorated the roof of Kakujo's 1993 Eurovan Westfalia with his artwork.

Hayashi's Vanagon impresses with its Westfalia roof and syncro all-wheel drive. He bought the VW from its second owner and is still

Everything packed?
Whenever possible, the friends leave Hamamatsu and look for that perfect wave on the Pacific

Japanese picnic. They do not need much to be happy. There is space for everyone under the bus canopies. Kakaju is the kitchen chef tonight

inspired by details such as the kitchen with refrigerator and sink and the bed.

Kei bought his vintage 1969 T2 Westfalia four years ago. With its weathered ivory-coloured paintwork and simple curtains, it is clearly the oldest Bulli in the convoy. Kei's VW may be rusty on the outside – the result of many surfing weekends – but on the inside everything is clean and sparkling. What's more, he has fitted the bus with a set of BF Goodrich off-road tyres, so that it won't get stuck in the sand.

In the evening, the chums put up plastic sheeting and light lanterns. The ocean plays the background music. Kakuju was once a chef and takes charge of the cooking. It's a cool April evening but the group of friends sit together in a circle, united in their passion for surfing and the odd drink or two warming them from the inside. It all feels good. Very good.

Into the waves with the morning sun

At sunrise, they all crawl out of their campers, keen to experience the first surf of the day. Naoya joins them with a two-tone T3 Caravelle painted in blue and navy: "I've been driving the bus for four years. I love its original character."

The men set off towards Nakatajima, one of the longest beaches in the district, which is famous for its dunes and the sea turtles that lay their eggs there. A number of Mercedes Transporters join the VW buses, and all of them drive down to the beach, causing the walkers there to stop and stare. One of the Bullis appears to have a slight problem. It's no surprise, given its age and a life full of salt and sand. But everyone lends a helping hand, and the camper bus is soon on the move again.

The Bulli has been a surfing icon since it first came onto the road, and it has a wonderful property. It acts like a magnet – on its owners, their friends, families and even strangers. The Bulli brings people together, helps to forge friendships and make discoveries. So they move on from beach to beach, the magnificent seven. Until work calls them back to their daily routine. But not for long; they're soon back on the road in their Bullis, back to the waves and the beach and to strangers who become friends. ❤

They tell stories, they listen, they dream. Surfing life feels good – even the dogs enjoy it

LOVE

JIMI WENT TO THE RAINBOW

Freedom, music and love. Barbara and Andreas Dünkel's Samba embodies the whole essence of the flower power years – but in its very own way

Jimi Hendrix smiles forever young. The wish for peace and love shines brightly in flowery colours and goes right to the heart. A yearning is felt. A yearning for past times, for the late 1960s when the flower power generation questioned traditional thinking.

Andreas and Barbara Dünkel's Bulli embodies all that is good and beautiful from these revolutionary years. The couple does not live in the past. They live life to the fullest. But they have never forgotten their roots – and they are hippies. Roots grow. And so do requirements. And this is why their hippy bus is so completely different. It confidently wears its high-gloss paintwork, it is not afraid of a little exaggeration, it is somehow modern, somehow ideal – like a perfect Hendrix concert. That never happened in reality, although every concert was perfect in the hearts of the fans.

That is precisely the essence of this bus that was delivered to the USA in the 1960s – in the form of the Samba special model that is especially coveted today, the model with its many windows and remarkably large sliding sunroof that once gave small tour operators a profitable income and large families some unforgettable holidays. It also gave the hippies an unhindered view of the stars.

Flower power perfection: Barabara Dünkel loves good design and lives VW Bulli – even in her choice of hand bag

In Woodstock, at the seaside in Goa or in the vast expanse of the American West. How much of this did this particular Samba bus experience? Andreas Dünkel doesn't know for sure. He only knows that he spent a long time looking for this bus.

Andreas is an avowed car and motorbike enthusiast. And he had wanted a genuine hippy bus for years. A Bulli to please his family and an eye-catching vehicle for his company, Motorworld.

A treasure awaited on the other side of the Atlantic

Andreas Dünkel had owned a classic Samba before, preserved in wonderful original condition. But repainting the patina bus would have been a big mistake in his opinion. This is why he started looking for a second Samba. He ultimately found one with the vehicle identification number 247 0459 96. It originally rolled off the production line in titian red (L 555) with a beige/grey roof (L 472). The bus must have done good service, because the old richness had long since faded, and the colours were barely discernible any more.

Dünkel brought his new precious possession across the Atlantic to restorer Kurt Reck in Hohentengen, a small town somewhere on the Rhine.

Despite everything, the Samba had come through the decades pretty well. The expert's eye saw that the poor exterior condition concealed surprisingly good inner substance. "Although the T1 was pretty worn out, it didn't have much rust", Kurt Reck says. "Furthermore, we discovered old accident damage on the left side." This is why the classic vehicle was first put on the straightening bench. The chassis was returned to its original shape, and the accident damage was discretely repaired. Then the technical

The destination is called Woodstock. It just needs to be fuelled up. The highway awaits the bus with the gentle eyes

components were reconstructed true to the original. Andreas Dünkel insisted that his hippy bus must be suitable for everyday use and should also be able to easily handle a journey of several months – at least theoretically. Goa still beckons. Who knows.

In close cooperation with Andreas Dünkel and a few VW bus specialists, Kurt Reck went looking for parts. And they were found: in Switzerland a store operated by the Swiss army was discovered where spare engines in original packaging were still waiting for use. Swords to ploughshares – the hippy community would have liked that. Now the Bulli engine is making love, not war.

The brand-new gearbox had a much longer journey: it came from Mexico. There were also many parts that could not be found anywhere, they had to be specially made. "Ultimately, restoring a vehicle that is half a century old is always a major job", Reck knows. But even the old Chinese book of proverbs, the Tao te Ching, which the hippies read enthusiastically, says that a long journey begins with the first step. The courage to go a long way and not be afraid to take up big challenges – these are without a doubt things that the hippies and the restorer have in common.

Jimi Hendrix laughs from the driver's side

As the restoration work in Hohententgen progressed step by step, the Dünkel family started discussing the interior and exterior design. This is where Barbara Dünkel, Andreas' wife, took the lead. She was uncompromising in the choice of motifs: "For me it was clear that Jimi Hendrix had to go on the bus – in a prominent position on the driver's side", she remembers. "We were also sure that we wanted all the slogans that defined our youth to be represented: All you need is Love. Flower Power. Make Love – not War."

Whilst the eyes around the front headlights were quickly made up in hippy style, the search for the right place for the Indian deity proved more difficult. It ultimately found an appropriate spot, crowning the front of the Bulli, right below the large VW emblem. The slogan Legalize Cannabis was positioned next to the left tail light. However, the most important thing for the flower power Bulli was its colour scheme. Here, an almost

ALL YOU NEED IS LOVE
AKE LOVE
NOT WAR
LOVE
MOTORWORLD

FLOWER POWER
ALL YOU NEED IS LOVE
PEACE LOVE
ROCK 'N' ROLL

Stillness. Just the bus and nature. A perfect reflection in the water. What could be more beautiful?

Type
T1
Model year
1966
Built in
Germany
Power (kW/hp)
32/44
Engine (displacement)
1.5 l boxer
Colour
Yellow/purple
Special features
Airbrush paint

Name
Andreas Dünkel
Year of birth
1964
Place of birth
Ulm, Germany
Place of residence
Schemmerhofen, Germany
Profession
Businessman
Passions
Classic cars

psychedelic mix of yellow and purple was finally chosen that made the body positively shine and also defined the look of the interior, which is painted in yellow with carefully coordinated violet and beige skai leather on the seat benches. This was a challenge that upholsterers do not get every day.

With a lot of time for every detail

As the technical aspects of the vehicle were being got into shape in Hohentengen, Bianca Hennig took care of the colour scheme and motifs. She works as an artist at a little place called Ofterdingen near Stuttgart and is extremely skilled with an airbrush gun. She had been well acquainted with the Dünkels for many years. "Bianca had already designed a few motorbikes for me – and perfectly", Barbara Dünkel remembers. "So, of course we also wanted her to apply her skills to our hippy bus."

The whole design process then took around six months. "But the results are great – it's just what we wanted."

Barbara and Andreas Dünkel haven't yet driven their bus to Goa, but they have been to several classic car events. And wherever they go, their Samba bus is a focus of attention, it is photographed, appraised and admired. Whilst the younger generation love the colours, the eyes of older admirers sometimes get that unmistakeable beaming look. A radiance that shines inwards and is reflected there in the eyes of Jimi Hendrix. ❤

Uncompromising. The exterior colours are repeated inside. Everything shines, is new and well thought-out. A VW with a beauty that is certainly unmatched by any previous VW

BUS

LIVED IN AND LOVED

Worn paint. Signs of age. Features that appeal today were frowned upon when Rikki James was doing up his 1954 Samba. But he felt it was the right thing to do

What is typical actually? Rikki James lives in a typical English brick-built house. The story of how he became a fan of Volkswagen vans is no less typical. Dad had one, the son grew up in it – and he then wanted one just like it for himself. Rikki's appearance is not typically English. But it is typical Bulli driver. Dreadlocks. No clichés, no posing. That's just the way Rikki James is. Someone searching for his own path without losing his way, someone for whom family is everything and who works hard to make time and space for them. Time and space that has always been enclosed by VW panelling, because his wife Vivienne and daughters Daisy and Lulu share his dream of life in the bus. It's not a case of hobby and family, but hobby within the family. Every day – four people and a Bulli. "My children have grown up with it. In the past, while we were still living in London, we had a VW 1600 Estate and this bus. For ten years, I drove to work every day with the Samba and this monster engine. Picked up the kids from kindergarten in it and used it to do the shopping. Summer and winter." Only the English can be so wonderfully eccentric. Nowhere else in the world is there so much affection for the VW bus as on the island. There is hardly a village that doesn't have at least one. "And if you do find such a village, there is at least one person there who knows someone else who owns one", says Rikki James, for whom a holiday flirt determined the rest of his life: "When I was six years old, I was allowed to go on holiday with friends. They had a brand-new 1966 T1 Camper. It was a tremendous difference to camping in a tent, which I was used to. In 1974, my father then bought a Bulli with a Devon conversion. When I was 18 and had passed my driving test, I borrowed the bus to go on holiday to Biarritz in France. That was where I saw a Samba for the first time. And that's where it took hold of me! Since then I've been hopelessly smitten by T1 Bullis!" Soon afterwards, the first bus of his own with a split front windscreen was standing outside the house. Rikki James began to immerse himself more and more in the subject, made the acquaintance of the Split Screen Van Club, or SSVC for short, in 1989 and has been heavily involved in the British Club ever since.

The James family loves the 1954 VW bus. When the bus door closes, daily routine stays outside. With Safari windows and the folding roof open, a holiday mood fills the interior

Eight small roof windows were missing

After a short while, the man with the dreadlocks realised that his bus was definitely missing eight small roof windows and a pop-top roof. He had to have a Samba! "I found a relevant ad in the SSVC magazine. Of course, I didn't have any money. Just my bus as something to barter with. I drove there,

"My first Samba didn't have any heating. In winter I always had to drive around wrapped up in boots, hat, gloves and woollen blankets!"

put my deal to the seller – and he actually agreed. He must have thought I was insane because his Samba was a heap of scrap whereas mine was in good condition. I've no idea what I was thinking at the time. In all likelihood, I wasn't thinking at all. I wanted the bus", explains Rikki James with a chuckle. "In the summer, my first Samba bus was like a dream, in winter dreadful. The bus had no heating, and I would always be driving around wrapped up in winter boots, a hat, gloves and a woollen blanket. It must have looked hilarious!" In 1992, it was in this Bulli that Viv and Rikki set off for the "greatest adventure of our lives. Apart from having children", the bus fan remarks with a grin. "We took the ferry from England to Santander in Spain, then to Portugal, back to Spain and on to France, where we attended the VW meeting in Menton. Italy was next on the agenda, and from Brindisi we continued by ferry to Greece. Then from island to island and back to the mainland. And suddenly the money had run out! Nonetheless, they were six fantastic months!" And Vivienne James adds: "When we arrived back home, we had covered exactly 10,067 miles. In a 1967 Samba!" Afterwards, the buses got older and older: vintage 1965, 1960, 1958 – and then finally this one from 1954. Rikki has a gleam in his eyes: "I simply had to have the bus. You don't get two bites at such an early, unmodified Samba!"

For daughters Daisy and Lulu there is no life without a bus. When they were younger they were picked up from kindergarten every day and went shopping in the bus

Dents and scratches tell of a tough life

Unlike the previous buses, Rikki James left the 1954 model as it was on the outside. "I'd had enough of restorations. The van fascinated me because it was so original on the outside. Original coat of paint. Dents and scratches as evidence of a hard life. In 1996, however, barely anyone understood that. The expression 'rat look' wasn't used in connection with classic buses at that time." However, as engine and gearbox already came from a newer model, the father of two had no qualms about making further changes. "I'd always had this vision of a very fast, very early bus. Travelling fast means spending less time on the motorway. Having more time at a meeting. Being able to spend more time at home. With the 1954 bus, I can move into the outside lane, put my foot down and tear down the motorway at 130, 140 km/h." With a 2.6-litre engine and just over 200 hp, that's eight times more than the original model. Plenty of power. Of course, that also calls for disc brakes all round, an improved chassis and a Porsche gearbox.

Almost every summer weekend, the James family are out and about in their Samba. "We're always on the move with our monster. At events across Europe or simply having a picnic locally", explains Vivienne James. And her husband adds: "As soon as we get into the Samba, we automatically switch into leisure mode. As if closing the bus door blocks out normal daily life." That may well be due to the many holiday memories. "Once or twice a year, we go off camping, and in 2012 we went on a road trip through the

13US. The registration number that looks like "BUS" is worth a small fortune. If Rikki James were to sell it, he would easily get a five-figure sum in pounds

Showing the traces of time. Rikki James likes the cockpit the way it is. It makes you wonder who might have first listened to the sounds from the now rare Telefunken radio

Type
T1
Model year
1954
Built in
Germany
Power (kW/hp)
over 147/200
Engine (displacement)
2.6 l boxer
Colour
Sealing wax red,
stone grey
Special features
American Racing Torque Thrust D magnesium wheels

Name
Rikki James
Year of birth
1960
Place of birth
London, Great Britain
Place of residence
Steventon, Oxfordshire, Great Britain
Passions
Good food, travel

Pyrenees with the children. When I told Lulu and Daisy that we wouldn't be flying off to the south for our summer holidays as usual, but would be taking a long bus trip, the two of them looked at me as if I were from another planet." The affable Englishman rocks with laughter. "They couldn't get their heads around that! After a few days on the road, though, their doubts changed to pure enthusiasm. A different place every day. Camping rough in the Pyrenees. Not a person in sight! No house, no TV, no Internet. Simply swing open the doors, set up the camping chairs, make something to eat, a glass of wine. The children on bikes, bathing in the lake. Life doesn't get any better, does it?" Rikki James continues animatedly: Viv and I would sleep on the bottom in the bus, the girls upstairs in the elevated roof, which looks like a roof box when it's folded up. A ladder gave them independent access to their bedroom, and a peaceful break from their parents. And we were able to sit cosily outside the bus and enjoy each other's company. An unforgettable trip!" "Since then, the girls keep asking when we're planning to go on the next road trip", Viv adds. And when will that be? "At the moment, work barely leaves me any spare time", Rikki answers. "My daily schedule has changed a lot since we've been living outside of London. We live in the country, and I'm faced with at least two hours' driving every workday." But at least it's in the VW bus, because there is of course a new T5 parked next to the 1954 Samba. "I've also had a Porsche and a BMW, but I prefer the Transporter to any saloon or sports car. I'm a bus person. Through and through!" ❤

Once the key is in the ignition, the adventure can begin. Vivienne and Rikki James have already travelled all over Europe. And always in the Samba van of course

SHAGGY AND THE MOUNTAIN MAN

Daniel Norris only follows rules in baseball. He is a millionaire and a pitcher for the Toronto Blue Jays. If his league bosses and fellow players had their way, he would be living in a villa. But to him money is simply too restrictive. He enjoys the freedom of living in his Westfalia camper

In professional baseball, very young men can become very rich very quickly. In 2011, Daniel Norris was just 18 years old, and he showed great talent as a High School pitcher. One morning he looked at his bank balance and saw a number in the millions. The Toronto Blue Jays had paid him two million dollars under a player's contract to pitch for their farm club.

Like many other baseball professionals who become instant millionaires, Norris treated himself to a German car. His fellow players would choose a Porsche 911 or Mercedes AMG. But Daniel Norris picked an old Volkswagen.

As a boy in Johnson City, Tennessee, he dreamed of owning a VW Westfalia camper, and now his dream has come true with a beige-coloured bus he bought for 10,000 dollars. Daniel named his VW "Shaggy" – the name of the sleuth from the Scooby Doo cartoon series.

Norris says that the camper helps him to keep his feet on the ground. And that isn't just a cliché for curious journalists who ask him about his "mountain man" lifestyle. Every year since signing with the Blue Jays, Norris has driven his Westfalia the 1,100 kilometres from Tennessee to the team's training camp in Dunedin, Florida. The reporters sensed the next big deal in baseball and made a big fuss. And this made the story of the young athlete's hippie-like lifestyle all the more interesting.

Photos of him with the camper gave him an "anti-star" image. Especially the photo in which he appears to be shaving his beard with an axe. Crazy but true: Norris achieved fame with his anti-star image.

But even now, he still has his wheels and his feet on the ground. While his team-mates took up residence in rented apartments, Norris camped in the VW and limited his monthly budget to 800 dollars. After training, he didn't sit down at a table in a restaurant. Instead he climbed into his Westfalia at his parking space at a Walmart where he fried a chicken fillet in a pan then sat down to read with a cup of home-brewed coffee.

Norris is thankful that he has come so far. He is young, and he follows his own path. Regardless of how high and far he goes in his career – "Shaggy" will always be there.

Daniel Norris accepts the hype about him. The bus gives Daniel Norris an escape from the star hype and a chance to value the simple things in life. In nature there are just him, the bus and his axe

Type T2
Model year 1978
Engine 1.6 l boxer, 37 kW/50 hp
Special features
Home on wheels

Name Daniel Norris
Year of birth 1993
Passions Baseball, living outdoors

DR.OETKER
HAUSFRAUENBERATUNG
DR.OETKER
HAUSFRAUEN
BERATUNG
BE 00 100

HAPPINESS IS THE BEST RECIPE

It's all Oetker here: the Bulli, its owner, the location – even the photographer

It's a foggy spring morning. The friendly face of the approaching T1 bus sparkles through all of the grey. Nina-Maria Oetker lifts up her camera. "He's coming!", she says. She is referring to Rolf Stövesand, whose life's work was to invent pizzas for Dr. Oetker, a leading company in the German food industry. Rolf's bus has a colour similar to that of vanilla pudding. The rumbling of the bus's boxer engine grows nearer, and Rolf waves. He is familiar with every inch of this place – the site of the Dr. Oetker factory in Bielefeld that is rich in tradition. The development department of the large-scale food manufacturer has relied on Rolf's ideas for some 30 years. The "oven-fresh" pizza he invented, for instance, is still a huge global bestseller today.

Rolf, a baker and pastry chef by training, steps out of the bus with a swing in his step: he's already reached retirement age – but you'd never know it, because his Bulli keeps him young. Naturally, he drives an Oetker Bulli; after all, Rolf Stövesand is an Oetker fan through and through. He's also a pizza fan. And a Bulli fan. Once, when Rolf was looking through the Dr. Oetker archives, he came across some photographs of old VW buses that the enthusiastic company had begun sending out on tour in 1950 – equipped with "kitchen helpers" to assist housewives at the dawn of Germany's Economic Miracle. One such helper was baking powder, the product that made the Dr. Oetker brand famous in its early days. Then there was pudding powder that made delicious desserts. Oetker cookbooks are also legendary – practically no household was without one back then. They were like a recipe for marital bliss, so to speak – and this is reflected in the words on Rolf's bus: "Good Advice for Housewives"; in the 1950s, such a well-intentioned slogan was not considered derogatory at all. "This is not an original Oetker Bus", Rolf points out. "It was previously a fire department vehicle." Rolf restored the T1 together with his brother, and today its paint finishing, lettering and graphics are all the same as those used for the originals. "Not all of its equipment is as it was in 1958", admits Rolf. For example, the bus has a 12-volt electrical system, the rear lights are larger units from a later model, and the engine has an output of 50 hp rather than 30. "For safety reasons", explains Rolf, who also likes to take his bus and Westfalia trailer on long trips to various festivals where he bakes and sells "Tarte flambée." Nina-Maria Oetker gives a thumbs-up. She is pleased with the photos she's taken. Our farewell with Rolf is brief but heartfelt. The T1 disappears back into the fog, heading out to a children's festival where it will provide some freshly baked treats. Rolf Stövesand and his Oetker Bus are sure to arrive on time.

Type T2
Model year 1958
Engine 1.6 l boxer, 37 kW/50 hp
Special features Former fire department vehicle, converted to a Dr. Oetker promotional bus using actual historical models as a basis

Name Rolf Stövesand
Year of birth 1947
Passions Inventing pizzas, restoring old Volkswagens, baking "Tarte flambée"

In front of the glass facade of the Dr. Oetker world is where Rolf Stövesand parks his Bulli He is a consultant for housewives. As he likes to say: "Starting at 9 o'clock at night."

ANYTHING BUT NORMAL

A plain T3 won't find its way into the garage of Dirk Klöß. Tipping, crossing rivers, camping, a touch of speed – his buses always have something rather special. And Dirk Klöß is not just any old bus fan, but a true mover and shaker in the German VW bus scene

Four T3 buses – and each has its very own history, has many special details and is always original Klöß

Cast off! The T3 as a master swimmer that can navigate on the high seas. Its nautical equipment is now historic

Type
T3
Model year
1983
Built in
Germany
Power (kW/hp)
92/125
Engine (displacement)
2.5 l
Colour
Dark blue
Special features
540,000 km (driven by himself) one-of-a-kind camper (self-built)

Name
Dirk Klöß
Year of birth
1966
Place of birth
Gießen, Germany
Place of residence
Gießen, Germany
Profession
Self-employed
Passions
VW bus! What else?

Everything under two metres forty is okay. Yet every extra millimetre would be too much for the holiday in Scandinavia. For then Dirk Klöss and his girlfriend would have to pay the expensive fare for the car ferry heading north. Pragmatic considerations such as this were important for the former student before he converted his two-tone 1983 Caravelle CL – with its 78 hp, only one owner and 6,000 kilometres on the clock – into a motor caravan. The elevated roof from Polyroof would meet the requirement – if Dirk were to lower the chassis a little as well. The plan worked, and the second career of the Caravelle as a camper bus was able to take off. But 30 years ago who could have imagined that it would last so long and still be running today.
That, in brief, is the story of the first bus that Dirk Klöss owned. Far in excess of half a million kilometres, over 30 years, accompanied by love and marriage. And yet this bus is more than just a phase. It is the most important bus in the Klöss garage, a type of big bang followed by a universe. For shortly after Dirk Klöss had bought the Caravelle, it became clear that the Volkswagen T3 bus – this transition from the air-cooled to the water-cooled engine world, this move away from smooth curves to practical square lines – was more than just a car or a hobby for Dirk.

Legend of the amphibious bus

Dirk Klöss received more and more enquiries from other T3 fans as to whether he could offer them advice or help – and this was how his own firm, which was dedicated almost exclusively to the VW bus, took off – almost on its own – while Dirk was still studying at the beginning of the 1990s. When you have a company like this, a small vehicle collection can't hurt. On some occasions, he didn't even have to look for the buses; they found him. One example is the amphibious bus, which he bought in 1998 from his friend Peter Hille, who is also the founder of the VW Bus Tuning Meeting at the Hockenheim Ring. An absolute one-off with a history that is still somewhat obscure. The bus was converted into an amphibious vehicle by Gerhard Thiry of Speyer, Germany in the early 1980s, and it was also spotted afloat in rivers on numerous occasions. Whether a scheduled trip to England via the English Channel ever took place or is just a myth remains unclear, even today. Dirk is still undecided as to whether he should leave the vehicle as it is or restore it. Restoration would raise the tricky question of what is really original and how to restore this condition, because some things have

Let's go swimming! With turn indicators from a T2 and a dolphin as a little mascot, this bus had already gone swimming in the English Channel. Or hadn't it?

PHOTOS THEODOR BARTH **TEXT** CHRISTIAN SCHLÜTER

ONTOUR-TEAM.de
Leben retten
MM 35521
VW der Schwimmeister
ONTOUR-TEAM.de

ONTOUR-TEAM.de
ONTOUR
Ich bin ein VW
GI DK 401

been changed over the course of time, and, what is more, the old vessel's engineering has become very difficult to obtain.

The King's double cab has six cylinders

Another fixed star in the Klöß universe also raises the issue of originality: enter the double cab from 1982, with a background story that could fill a novel. That is because the vehicle was said to be built at the request of the King of Spain. It was never actually delivered, however, and thus became a symbol of innovation for the Oettinger/Okrasa company. In 1982, the vehicle was first registered. In 1983, it was given an impressive two-tone paint finish in gold and blue and a 3.2-litre six-cylinder water-cooled boxer engine. At the time, this engine was developed almost to production maturity and, shortly before completion, it was sold as a complete project to the Oettinger company, who then marketed it. Up to 1987, the vehicle repeatedly underwent conversion and was exhibited for sale at trade fairs. Even back in 1983, it was priced at 149,500 Deutschmarks (including TÜV certification). That was an incredible price at the time. A standard double cab was already available for less than 23,000 Deutschmarks. Before the vehicle reached Dirk, it had been modified a number of times and painted

Always on the ball: Dirk Klöß is a mechanic by passion, someone who wants answers to his technical questions, and finds them

To restore? Or preserve its condition? Dirk Klöß' buses always raise these questions. There is no simple answer. You just have to feel the answer

in the Orly Blue of the Last Limited Edition. In the case of this vehicle, too, restoration would raise the question as to which state it should be restored to? The standard production vehicle in red, the gold and blue of the concept or the Orly Blue from the 1990s? If it were to be restored, Dirk would definitely do the work himself. In the meantime, the restoration manager has become famous for his high level of VW bus expertise far beyond the borders of Germany. Dirk is a member of the VW bus scene and is always a sought-after discussion partner at meetings. His customers see him as a custodian of the bus philosophy. He takes care of his customers' "family members" as if they were his own, and you only entrust them to someone you can depend on. He doesn't talk about some of the prestigious names amongst his customers; discretion is a point of honour. He also subscribes to another point of honour: a member of the family always remains a member of the family! Anyone who belongs to the Klöß clan can stay as long as he likes. The only exception so far has been a camper bus that Dirk donated to the charitable association "Lebenshilfe Giessen", which raffled it off to raise funds to support the disabled. All the other buses are still there or have been added since – such as the Multivan Magnum with a six-cylinder engine. Dirk's 1989 version has an engine that produces 240 hp. A very rare engine that was used in the Porsche 911 for only a short while – and which is therefore all the more scarce in the buses.

At first glance, the red tipper from 1990 is the only one amongst all the vehicles that doesn't really fit into the family portrait. It's a bit like the weird uncle who still lives in the country while all the smart brothers and cousins have been experiencing the wide world. Yet there is one good reason why the tipper is part of the family: whilst his schoolmates were always driven to school in the same soulless SUVs or sports cars, Dirk Klöß' offspring asked his dad to give him a lift in the red tipper. After all, who has a tipper? This seems to have firmly rooted the desire for unusual vehicles in the family. And dad's collection would seem to provide precisely the right foundation.

MM C 424

THE WILD LIFE

Driving along the legendary hippy trail with your own Bulli to Kathmandu and Goa – many people would like to have done that. Jürgen Schultz did it

"You can run into big problems if you hit a cow with a car in India." It all sounds so normal when Jürgen Schultz tells the story. Still, he ought to know, since it actually happened to him. "Luckily, we were able to ransom ourselves with a bottle of whisky", he says, "otherwise it could have ended badly – and we wanted to head on to Goa, after all!" It seems that those who wish to see the world also have to take some risks. "But we were completely unaware of such risks", says Schultz, originally from Memmingen, Germany. When you see the twinkle in his eye, you have to believe him – even today, years later.

His first trip to India was marked by a type of awed naivety. "A friend and I simply wanted to find out how far we could get with a 1959 Beetle that had cost 100 Deutschmarks", he explains. After completing their preparations by mounting a roof rack, they were ready to head off to New Delhi. "We came back six months later and junked the Beetle", says Schultz, a pleasant looking man with a beard. His heart still bleeds for the car, though. "But that's what we used to do with old jalopies", he explains.

Schultz's blue 1956 T1 is no jalopy, but it is old. The vehicle is a genuine marvel, as it has remained in great shape after many long trips, including its most recent one ("just to Afghanistan") in the autumn of 2000. The bus came back complete with a cashmere blanket on the bed, a "left-hand drive" warning sign from Sri Lanka on the back and a hodgepodge of spare parts stored in the double hinged doors. The vehicle not only exudes history; it IS history.

And Jürgen Schultz is the one who experienced it. Altogether he has owned three T1 models and a T2a, as well as various Beetles. At the moment, he drives a 1956 Karmann. "Still, the Bulli remains the true car for a hippie", says Schultz, who's been to more than 80 countries. "You see, if you had a VW bus back in the day, you weren't just a hippie, you were a super-hippie", says the well travelled southern German with a grin.

Back again. Arriving home after many journeys.

Every millimetre of sheet metal filled with life, love and karma.

A wonder that these buses still exist

From government bus to hippy mobile

It's probably as broad a smile as the one he had when he first bought the bus. "I saw it on the side of the road and made a sign with my hand to make an offer", he recalls. "It actually worked." The previous owner had converted the bus from a government vehicle into a camper van but was now looking to get rid of it. It therefore perhaps wasn't necessary for Schultz to dress up nicely and trim his beard like he did in order to make a good impression when the deal was finalised. After all, even his explanation that he wanted to

Jürgen Schultz was there. And the map shows the route he took. Its time was not measured in weeks or months, but rather in moons. Somehow time ceased to play a role at some point

Type
T1
Model year
1956
Built in
Germany
Power (kW/hp)
25/34
Engine (displacement)
1.2 l boxer
Colour
Blue (currently)
Special features
Ex-government vehicle of the German Postal Service, self-made camper van conversion, originally preserved "Hippy bus".
Today, the vehicle is located in the "Erwin Hymer Museum" in Bad Waldsee, Germany

Name
Jürgen Schultz
Year of birth
1942
Place of birth
Memmingen, Germany
Place of residence
Memmingen, Germany
Profession
Retired
Passions
Air-cooled VWs, travelling, languages

The folding door leads into another world, pulls you in to Goa. This bus might be ready to go tomorrow. Are you coming along for the trip

drive the bus to India with his brother Bruno to join the "drop-outs" there didn't seem to shock the elderly owner. A short time later, in September 1977, he did in fact head off to Goa – the mecca for real hippies – in the bus, which is on its third engine today. This was Schultz's second big trip to India with a car; the first was in 1973–74, also with a Bulli, which he ended up selling in Nepal for the equivalent of 1,000 euros.

He has remained committed to his 1956 model, however, even if everything takes a little longer with a 34-hp engine. "In Afghanistan, I once had to drive up a pass in reverse for three kilometres because the bus couldn't handle the incline in first gear", he recalls. Petrol was sometimes paid for by "petrol payers" – hippies with no bus of their own who hitched rides for days or even weeks on the way east to the tropical destination of Goa.

"Everybody was smoking marijuana everywhere, of course", Schultz says, "and I can tell you I tried it as well. Still, we weren't at all interested in getting high all the time, so we just stopped doing it." Instead, we enjoyed coconut cocktails. "Those parties under the full moon on the beaches of India were legendary", Schultz muses somewhat wistfully. "The people, the stars in the sky, the easy life – it was like a dream. We were part of something big, and we fully enjoyed every minute of it."
Schultz's interest in travelling began while completing his military service. "The army sent me to Sardinia and I really liked it there", admits the greying old hippie. "I could never get rid of the travel bug after that…"

After he was discharged from the army, he took advantage of a programme for free flights to Johannesburg being offered by the South African government: "They were trying to attract whites to the country, so a lot of people used the free flights for a holiday and then simply left when they were done." Schultz stayed, however. He went to work in a steel mill and later got a job as a salesman for VW parts. "My boss there was from Leutkirch in the Allgäu region of Germany", he explains. "That's near where I come from. So the interview was sort of a 'down-home' thing." A few years later he travelled to Paris: "I also worked as a parts salesman there for a few lousy francs that were barely enough to pay the rent. It was the language I was interested in — languages are everything to me!"

After returning to Germany, Schultz got a job at MAHAG, a well known VW dealership in Munich. He still had the travel bug, and his boss let him take a one-year leave of absence on three different occasions: "I did it as unpaid holiday, but later I simply quit – you just have to go out and do what you want!" One thing he never wanted to do was get married. "And I still don't regret that decision", he says. Not even an offer from the Police Chief of Sri Lanka could persuade him otherwise: "He wanted to trade his daughter for my bus …"

Quit your job, drive away. It's that easy!

Schultz made two trips to India with his Bulli in the 1970s: one in 1973–

Everything is still there. Books, maps, lucky charms and a small kitchen unit. It doesn't take much to be happy. Just the courage to let go

Today, the hippy bus is in the Erwin Hymer museum. Here, we see the bus just like it is. Its age, dignity, history - this Bulli has it all

74, and one from September 1977 to September 1978 in the '56 Bulli he still has today. Jürgen Schultz stresses time and again how much his journeys helped him as a person: "You are constantly expanding your horizons, and you become more satisfied with what you have, even if it isn't much." Materialism or career aspirations were never his thing anyway. "When I was a young bloke, I had to do what other people said six days a week, and at some point I began wondering if I was crazy", is how he describes his job experiences in the '70s.

Even with all his trips from "here and now" to "there and then", Schultz has remained true to his nature in a most charming manner. "Oh boy – I was always a frugal Swabian", he says in his local dialect. That's why all his trips had to be cheap, and he also took photos with a used 1942 Leica camera. "Cheap, cheap – that's the ticket!", he says. It definitely also helped that the '56 Bulli he paid 1,500 Deutschmarks for hardly needed any fixing on the 40,000-kilometre trips to Goa and back. "It's simply a reliable vehicle, my Bulli", he says. That is why the thought of selling it never entered his mind.

Nonetheless, at some point the barn in which he does his tinkering had to be cleared out, so he approached the Erwin Hymer Museum in nearby Bad Waldsee and offered to loan the bus to them. The museum was thrilled to have it. "We were really happy to get it, because we were looking for just such a vehicle, and the pictures Mr Schultz sent us looked very good", says Susanne Hinzen from the Management Board of the Erwin Hymer Foundation, which operates the Erwin Hymer Museum. "Still, we didn't realise just how unique the bus was."

The hippie bus joined the museum's collection in 2014. It hasn't been altered in any way, and there are no plans to restore it. "We're displaying the bus as it is, with all of the interesting things Mr Schultz brought back from his last trip", Hinzen assures us, adding that she never had any reservations about including the highly modified vehicle in the museum's collection: "We want to keep the vehicles we have as authentic as possible; Mr Schultz's T1 has no need to be embarrassed; it exhibits the signs of the times it was used in, so to speak."

Some say that people come to resemble their cars a little over time. That's definitely the case with Jürgen Schultz and his Bulli, as both have had a lot of experiences that have left their mark on them. They both have a lot to tell as well, but neither seems to have grown up, even after all these years – and there's no reason to believe that they ever will.

The morning fog lifts and lets the sun shine through softly. Are we in India or southern Germany? Another look: it must be India. That's where the bus is

PRURIENT PROVOCATION

Ed Fox's 1959 double-cab pickup is a work of art, a sculpture on the asphalt. A vehicle that splits opinion. Love it or leave it. There is nothing in between

Robbed, but not giving up

The VW pickup, with its metallic gleam, rolls slowly into place in the shadows of the interstate bridge in southern California. Here it is still cool. But the water in the shallow puddles will evaporate by midday at the latest. The sound of the turbo engine reverberates from the ceiling of the concrete structure like an evil growl. A bit more manoeuvring, then the double cab – "Doka" for short – is positioned as Ed Fox wants it. Lights on. Like a fierce animal, the 1959 Bulli now looks down the lens. Photography professional Fox presses his Canon's shutter button. "Perfect!" A long time passed until the Hollywood photographer and car enthusiast was pleased with the look of his pickup. "Actually, you're never finished. But quite a bit has certainly been done since 1992", grins the Bulli fan. It was back then, on 23rd April 1992 to be exact, that the liaison began. "I was working one day with my father in the workshop. We were in the middle of tinkering away on a 1967 double cab, when an older gentleman turned up, watched us for a while and then said that he had a pickup just like that in his garage, should we be interested in another one. We were. I was! It cost me 1,600 dollars. Quite expensive back then, but it did come complete with Safari windows at the front." Over the next few years, Ed Fox experimented with colours and different parts, such as wheels and bumpers, in an effort to make this VW Bulli his VW Bulli. He was twice interrupted in the process by thieves. First, some youths tried to push start the Volkswagen to then make off with it. That failed, because the Californian noticed the attempted theft and was able to chase off the would-be gangsters. As it rolled away down a hill, he also managed to stop it from crashing into a wall at the last moment. On the second occasion the pickup disappeared. Just like that. That was in 1996. Ed Fox didn't give up; he found parts of his vehicle being offered for sale and put the police onto the thieves' trail. Dismantled but not damaged, he got his beloved double cab back. "When I put it back together, I got bare metallic hubcaps for it. And that got me thinking about stripping all of the paint off the entire pickup!"

Provocation and popularity

While the graduate of the Art Center College of Design was able to spend a great many hours on the Bulli's look at first, his focus moved on to other things starting in 1997. That was not due to any lack of motivation, but to his career, which was starting to take off. He began taking photographs for the mildly erotic 'Leg Show Magazine' and had an opportunity to combine

Function follows form. In Ed Fox's world at least. The visual appearance of this double cab pickup is more important to the Californian photographer than being able to change a rear wheel quickly

PHOTOS ED FOX **TEXT** GEORG OTTO

Ed Fox calls it "The Instrument". A clock on a wall in a shopping mall inspired him to make this multi-function display unit. The bracket came from the insides of the erstwhile speedometer

a very special predisposition with his work. "When I was 15, I noticed my predilection for female feet. I found that extremely embarrassing and didn't talk to anyone about it. Even when I was working for Leg Show, I kept a very low profile. But ever since the publication of my first book, 'Glamour from the Ground Up', even if I was in fact a bit uncomfortable about it before, I became much more relaxed about it!" Other forms of erotic play do nothing for the 48 year old personally. But they are certainly part of his professional repertoire. "In 1998, when my Doka was finished in its new, naked outfit, the world-famous VW Classic Car meet in Irvine was coming up. I was looking for something to provoke a reaction. So I had two young women chained bondage style to the tarpaulin hoops and drove like that onto the showground. The organisers didn't find it funny and I was given a warning. The following year they actually threw me out because of it", reports Ed Fox with a laugh. "And yet they had bikini contests there! What's

Type
T1
Model year
1959
Built in
Germany
Power (kW/hp)
over 147/200
Engine (displacement)
2.2 l boxer
Colour
None
Special features
Body only has clear coat

Name
Ed Fox
Year of birth
1967
Place of birth
Glendale, USA
Place of residence
Los Angeles, USA
Profession
Photographer
Passions
My double cab, my children, erotic photography

Many have already modelled for Ed Fox in front of the camera. Not just his pick-up, but also models such as Dita Von Teese, Ginger Jolie, Penny Flame, Tori Black and Belladonna. Although he has a good reputation in a not very reputable industry, he aims to quit erotic photography

the difference?" But with a dismissive wave he adds: "I didn't mind, putting on that show was worth it!" Since then Ed Fox's double cab pickup has become known all around the scene of VW cars with air-cooled engines. Fan magazines from England, France and Germany hurled themselves at the enfant terrible in print and his transporter, which was finished with just a single-layer of clear coat. Erotic magazines, too, discovered the pickup as a backdrop for photo-shoots. "That is how I brought my two great passions together", explains the American.

Reorientation

As he became better known, his circle of clients became broader as well. Gerber, Mario Lopez Underwear, Getty Images and even Disney are now clients of his. "I actually want to step away from erotic photography. And instead concentrate more on my – let's say – mainstream-compatible career." He made this decision partly for the sake of his two young daughters. "As a father, you set different priorities", says the man, who you might imagine to be covered in tattoos and perforated with piercings. In fact, Ed Fox wears shirts and has a trim short haircut. A good family man, who puts wife and children above all else, but who will nonetheless never give up his Volkswagen. "I have to admit that my wife would prefer to have me sell it, and my mother will probably keep on asking when it's finally going to get painted. But neither one nor the other is going to happen! The vehicle is part of me." One of the latest modifications is "The Instrument". That's what Ed Fox calls the add-on fitting positioned behind the windscreen's centre upright. He worked on its development for four years. "A clock on a wall in a shopping mall gave me the inspiration for it. I brooded for ages over drawings, clay models and wristwatch illustrations until I was finally happy

with the results." And then he still needs to find a solution for changing the tyres. "At the moment, due to it being lowered so much, it works like this: let air out of the tyre, undo bumper, remove front axle from spring plate, unscrew wheel bolts, take wheel off." He ponders briefly. "But perhaps that's also how it should be. Because form is actually more important to me than function. My pickup is after all more a work of art than a vehicle." ❤

BUILT ON WINE

Anyone can buy a camper van. But to swap 800 bottles of Mosel wine for a Bulli – to do that you need to be a winemaker. Or the son of one, like Tobias Stadtfeld

The Mosel Valley in the west of Germany is regarded as one of the very best regions for making wine. Along the meandering course of the river the grapes grow in the bright sunlight. With its many hills, the Mosel Valley is also an ideal training ground for ambitious bicycle racing enthusiasts and runners, like Tobias Stadtfeld, who grew up here as the son of a vintner. His passion: duathlon, at competition level. That means running, cycling and running again. But he also has a second passion: his VW camper bus, a T2a from 1971. Life without this Bulli is something the paediatrician can no longer imagine. "Initially, it was my childhood dream to own a camper van one day. And fairly quickly, I was also able to get my girlfriend at that time, who is now my wife Nina, enthusiastic about the idea. So, in 2010 I decided to invest the necessary capital and to buy a second-generation Transporter!" The side-windowed camper van was up for sale at the time in Holland, so Tobias, Nina and father-in-law Helmut Hard went to have a look at it. By applying his commercial talent, Helmut was able to knock the price down and – Tobias could hardly believe his ears – negotiated a very unusual part-payment. Two weeks later, father and son Stadtfeld set off for Holland to pick up the sierra yellow beauty, armed with cash and 800 bottles of wine. Thus, Tobias was able to fulfil his childhood Volkswagen dream. A camper van of this sort is a good deal. After Tobias helps children to get well all day, in the evening the camper helps him to relax. And once even helped him to get married. "I proposed to Nina in the camper on the beach in Holland. The wedding was due to take place in our dream country, Norway. With the camper, of course!" It didn't let the bridal couple down, though it was a close call. Just before the marriage ceremony, the petrol pump gave out and the bride had to push. "A picture that many wedding photographers would love to stage! In our case it became a reality!" Whenever he thinks back, Tobias still has to laugh about it.

The Norwegian tour, however, still took place as planned and forged an even closer bond between the young couple. "To me, travelling with the camper is pure freedom! Deciding for yourself where to go!" Tobias contemplates briefly: "It's captured very well in a quote from 'The Old, the Young and the Sea', a documentary film about the European surfing scene and the VW camper vans they had. It says: 'When you buy a camper, it changes your life. You become free as a bird: today you're here – and tomorrow somewhere else altogether. The world becomes your home.' That says it all."

Type T2
Model year 1971
Engine 1.6 l boxer, 32 kW/47 hp
Special features Modified interior renovation with sleeping bench and folding table, front end of body slightly lower

Name Tobias Stadtfeld
Year of birth 1981
Passions Air-cooled Volkswagens, duathlon, bicycle riding, competitive running

Paediatrician Tobias Stadtfeld frequently helps his father in the vineyard – even to this day. In return, Stadtfeld senior was happy to help him with 800 bottles of wine as part payment for his VW camper van

PHOTOS SVEN ZIMMERMANN **TEXT** GEORG OTTO

THE FIERY RED MACHINE

Extraordinary race cars require extraordinary race-car transport vehicles – like the T1 flatbed bus from State-of-Art

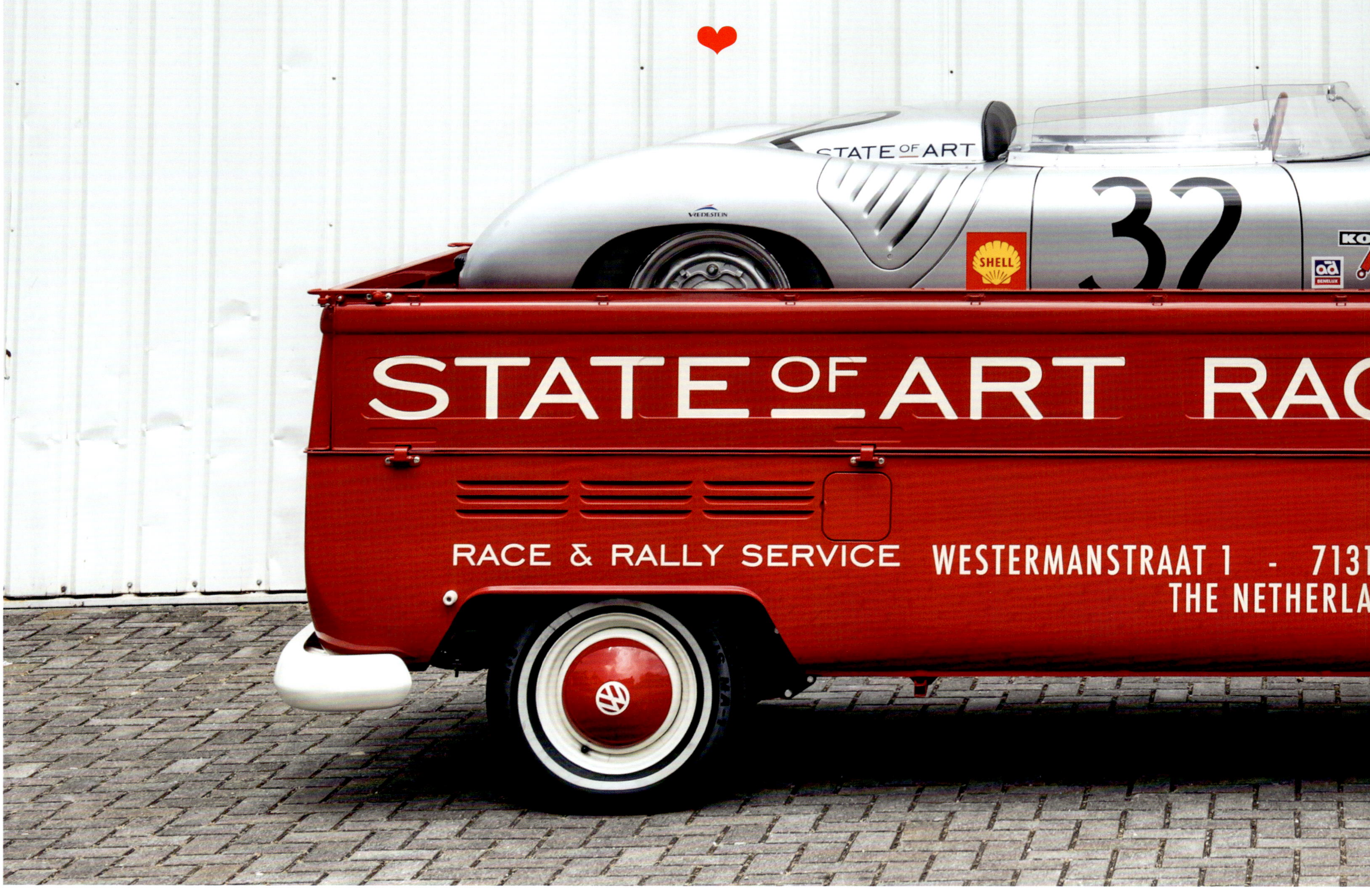

Around one metre 40 cm longer than a T1 pickup. Plenty of reinforcements were added – those are the parameters for a State-of-Art race team bus

32
STATE OF ART
32
32
SHELL
KONI
TATE OF ART RACING TEA
ACE & RALLY SERVICE
WESTERMANSTRAAT 1 - 7131 EH LICHTENVOORDE
THE NETHERLANDS
ALBERT WESTERMAN B.V.
PORSCHE

Albert Westerman is a fan of classic Porsche and Volkswagen cars. And he's a little crazy. Fortunately!

Type
T1
Model year
1967 (for the most part)
Built in
Germany & Netherlands
Power (kW/hp)
66/90
Engine (displacement)
1.6 l boxer
Colour
Polo red (like the historic Porsche race transporters)
Special features
VW T1 flat bed with cargo bed lengthened by 1.4 metres to transport a Porsche RSK; retrofitted based on the example of the historic Rennstall Bunker transporter

Name
Albert Westerman
Year of birth
1943
Place of birth
Lichtenvoorde, Netherlands
Place of residence
Lichtenvoorde, Netherlands
Profession
Entrepreneur
Passions
Porsche, VW, women and beautiful clothes

Wow, is that long! – Did Volkswagen build it like that? – This is how conversations often begin with Frank (Franky) Slenders who hails from the tranquil city of Bergeijk in the Netherlands. That's because people who catch a glimpse of the "State-of-Art" flatbed race-car transport bus he built can hardly believe that Volkswagen didn't manufacture the vehicle itself. Franky has a workshop near Eindhoven in which he restores vintage vehicles, many of them Porsche and Volkswagen models. His goal is to always come as close to the perfection achieved in Wolfsburg and Hannover as possible. Another of his goals is to bring crazy ideas to life. Strictly speaking, however, the race-car transporter wasn't his idea. It actually existed a long, long time ago. Not even the most die-hard auto racing fans today are familiar with name of Arthur S. Bunker Jr. from Kansas City, Missouri, in the USA. However, vintage car fans will probably know about the "Rennstall Bunker". That name refers to a Sahara-beige T1 pickup bus with a noticeably elongated wheelbase that was used by the "Rennstall Bunker" racing team to transport its Porsche 550 Spyder. However, the bus only really became famous after it was produced as a miniature model.

Still, not everyone is satisfied with a miniature 1:43-scale Rennstall Bunker. "I saw the miniature at a vintage car show and immediately realised it would be great to build a full-sized one", Franky recalls. "A friend and I often wondered who might be interested in something like that – and then it came to us: Albert Westerman!"

Cut to Lichtenvorde, in the northern part of the Netherlands. Enter Albert Westerman. It wouldn't be stretching the truth to call the sporty, tan, slim, grey-haired fashion apparel entrepreneur a "car freak". However, such a description isn't entirely accurate: Westerman is not some eccentric who collects precious treasures that he hides away in a garage and watches the layers of dust collect on them like the rings in a tree. He just happens to love Porsche vehicles like the 356, 911 and 550 Spyder – and he's got a lot of them. But they don't just sit around in the showrooms for his State-of-Art fashion label. Instead, they "earn their keep" in an enjoyable way by serving as promotional vehicles at vintage sports car races, tours and meets. Westerman's museum on wheels is on the road a lot, and in the car scene his State-of-Art label is highly appreciated for its Porsche and Mille Miglia fashion collections.

A guy like Franky Slenders never gives up

Now back to Bergeijk, where Franky and his team came up with a proposal that Westerman simply couldn't resist: the businessman with a big heart for crazy ideas gave them the green light. "But first we had to find a suitable bus that was in good shape", Franky explains with a sigh. Indeed, the automobile specialist looked at ten d fferent candidates, several turned out to be imposters. "In the end, we took the T1 pickup we had seen first, and which we also knew was still available", Franky says.

That's because although the bus in question was originally from the USA and had only 58,000 miles on the clock, it also had a completely wrecked roof. "It must have flipped over before it was re-exported back to Europe, which would also explain the low mileage", Franky says. "Nevertheless, its

A penetrating perspective: despite all of its reinforcements, the T1 still preserved enough space for a lower storage compartment named Vault

new Dutch owner still drove it for six years", he adds with amazement. "One day, after we bought it, the previous owner phoned us. He was stunned, he said, because someone had just offered him a roof that would have fit the bus. But it was too late, of course, and after we told him what we had planned with the VW and who we were going to give it to, he felt better because he had heard of Albert Westerman and thought the idea was great."

Franky and his team were now ready to get started. First, they fitted the bus with a new replacement roof. Then the team hacksawed the T1 into pieces. Franky built suitable side and cross members and converted the hollow-profile structural members that were over two metres in length into solid members "in order to make everything more rigid", as the perfectionist explains. "That was a huge amount of work that also required a great deal of precision. If you're off by even one or two millimetres at the start, you'll end up with big problems later!"

Franky ultimately lengthened the 1967 T1 by 1.4 metres, which ensured that not only a 550 Spyder but also an RSK or a Porsche 356 would fit on the flatbed if necessary. A metalworker friend of Franky's made sure the body panels were exactly the right shape. At some point, after many hours of work, the body shell was ready for sanding and sealing with epoxy resin. There wasn't much time left until the big "Techno Classica" vintage show in Germany. "We had promised that we would be finished by then, and we were hell-bent to deliver on that promise", Franky explains. They were true to their word, but first they had to make some technical adjustments. The 90-hp engine for their project was taken from a Porsche 912 and then modified at the State-of-Art workshop. The control cables for the throttle, clutch and brakes weren't much of a problem. For the cable harness, Franky procured a component similar to the original one and integrated it into the original harness. The team decided to use a Polo-red paint very similar to the one formerly used by the Porsche factory team for its own race-car transport vehicles. "We didn't want a complete Bunker reproduction, but our own distinct look instead", Franky says.

Westerman simply loves it. No matter where he pulls up with his spectacular T1 pickup with a 550 Spyder in the back, he is assured plenty of admiring looks. That's perfect, since the bus is primarily meant to advertise his fashion label. It's more than that now, however; it's a part of his beloved collection of vintage automobiles – and one that definitely makes it fun tc drive to the raceway.

90 hp at the rear just like a fine Porsche. The fine manners of a 356 are conjured from a Beetle engine, and the interior expresses an austere charm of those times. Nothing to distract, not even a radio

ON A MAJOR TOUR WITH LILLIE AND HARVEY

From the maternity unit into the midst of life. From a secure job to an adventure. All in the Bulli. Erik Hormann has some kind of primal trust in his Volkswagen. And rightly so

Love doesn't make you blind. Love opens your eyes to new things. And if love lasts a lifetime, then it will overcome all doubts. It was out of love for the Volkswagen bus that Erik Hormann founded his company Vantigo – which offers tours of a special kind in and around San Francisco. To very special places and on special occasions. All in a VW bus of course. And there's a particular reason for this: Erik's intimate connection with Volkswagens can be traced back to his birth. Before him his parents were VW bus fans and made their first trips in a camper bus. So it's no surprise that Erik and his twin brother made the very first journey of their lives – from the maternity unit to their home – in a VW bus. More precisely: in a blue-and-white 1971 Bulli. Later, the young Hormann family travelled around in a 1974 orange camper. They named it "Pumpkin". "My brother and I thought it was great climbing on to the mattress in the pop top," said Erik. "It was our mobile playroom."

As Erik's family was visiting the sights in their VW, his future wife Amy spent her childhood travelling around California in a 1971 Westfalia called "Root Beer". Like the Hormann family's Pumpkin, the root beer bus was named after its colour. Amy's father chauffeured his family on camping trips and to San Francisco 49ers (American) football games in the brown bus with a white roof.

Love discovered – for one another and the Bulli

Amy and Erik met, and not only did they fall in love, but they also discovered another love they shared. "When we met we soon found out that we'd both grown up with Bullis. For us that was a happy coincidence." The Bulli is the lucky star watching over their relationship – Erik is certain of that. Erik had already owned a few VWs before he and Amy exchanged rings: a souped-up dune buggy, a Type 3 fastback, three estates and a handful of Bullis. He would swap old video games for parts, tinker around and get advice about repairs and restorations from his father or from the internet. The couple moved from Chico in central California to San Francisco, and they quickly found out that owning a classic vehicle in the coastal city is a real challenge. Apartments rarely have their own garages, and can be costly to rent. That is why Amy suggested to Erik that they should use a VW bus that he was intending to restore for more than weekend excursions. Instead of having a VW bus that's fun but costs money, why not have one and earn money while having fun? "It was clear to us

Just come along, climb in, go on a tour with me!

Amy and Erik organise day trips and short journeys. Always in classic Volkswagen buses

Enjoyment is the goal on beer tasting tours. Discovering something new, the good sides of life. How could it be any better than in a Volkswagen?

that we needed to do something with our Bulli," Erik explained. And because he wanted to be his own boss anyway, he and Amy started up an agency for Bulli tours: Vantigo.

Erik used the bonus cheque from his secure full-time job in a technology company to buy a 1971 Transporter Deluxe, and he spent the next six months refurbishing the interior, hatching ideas for tours and dealing with pesky red tape. In August 2013, Vantigo offered its first VW Bulli tour. "I completely immersed myself in the project, spent nights, days off and weekends on tours and photo shoots alongside my job. Seven days a week. But I wanted our idea to work", said Erik. In March 2014, Erik quit his job to run Vantigo. Six months later he had already employed his first staff member. Today, Vantigo offers adventure tours, brewery or wine tours and culinary tours. Amy and Erik organise weddings, photo shoots and tailor-made events along the coastal Highway 1 and in San Francisco, Sonoma and Tomales Bay.

Customers can choose from two camper buses. There's Lillie: the 1971 Transporter Deluxe in Chianti red, pastel white and alabaster – with which everything started. Lillie is named after Lillie Hitchcock Coit, San Francisco's famous volunteer firefighter and benefactor of the Coit Tower. Lillie is backed up by Harvey, a Transporter Deluxe of the same age in Niagara blue, pastel white and alabaster – named after the progressive leader Harvey Milk. Erik is Lillie's third owner. He found her in southern California with 96,560 kilometres on the clock. The interiors of Lillie and Harvey were carefully reconditioned for greater safety and comfort. Cup holders have now been installed for the beer tours, and the carpets have given way to hard-wood floors. There is only one word to describe

Type
T2
Model year
1971
Built in
Germany
Power (kW/hp)
37/50
Engine (displacement)
1.6 l boxer
Colour
Chianti red, pastel white
Special features
The bus's name is Lillie, named after Lillie Hitchcock Coit, benefactor of the Coit fire tower in San Francisco

Name
Erik Hormann
Year of birth
1983
Place of birth
Chico, USA
Place of residence
San Francisco, USA
Profession
Owner of the Vantigo company
Passions
To show people the beauty of California through the windows of a VW bus

Stepping out and looking good. Whether at the ocean or in the city. Lillie, the bus, takes every guest to their destination. Looking good. But no one wants to step out of the bus any more

the engines and the characters of the two Bullis: authentic. "I haven't done anything that would disappoint a VW purist," Erik assures. "It's a perennial conflict, but if a change makes things more comfortable and everything a bit more contemporary, I don't have a problem with it."

Many of Vantigo's guests grew up with Volkswagen and enjoy telling stories about their childhood travels or tours when they were students. People from Chile, Canada, India, New Zealand and the world over chat with Erik about the incomparable VW Bulli. "Some buy their VW in Europe and explore the whole continent in it before they bring the vehicle back home. Others use the camper as a school bus, tool truck or flower bus. But they all get in with a smile. These Bullis are pure happiness." For some people, driving Lillie or Harvey on the famously steep streets of San Francisco might seem an unreasonable challenge. But Erik enjoys exploring his city in the Volkswagen. The wheelbase is shorter than many modern cars, enabling the Bullis to easily wind their way through the city's narrow streets. The guests are always pleased with the high level of comfort and the large windows in the VW. "With their tight transmission ratios, my buses are born hill climbers. What's more, the large windows and panoramic view mean that the people don't get travel sick. You don't go too fast – the drive is a sheer pleasure", said Erik enthusiastically. He particularly likes going up the Twin Peaks and travelling to the Sutro Baths or Fort Point. Erik is now restoring the third VW, a 1972 Deluxe with a sun roof, which he wants to imbue with a few special features and of course a name inspired by San Francisco. Erik's twin brother Hans is about to open his own office for Bulli tours in Venture – making the whole thing "a family affair", as Erik puts it. "And the Bullis are part of the family. If they get sick, you worry and give them extra special care. If they're well, you're happy, treat them well and give them a nice wash by hand. Harvey and Lillie even decorate our Christmas postcards. They are family and a whole lot of fun – every single day."

ON TOUR THE TOUGH WAY

Making your girlfriend get out on a hillside, changing a gearbox on a mountain pass and driving through holes metres deep in mud – that's the Shasta Snow Trip. And up to 50 T1 Volkswagens take part in it every year

He's going to make it! Even when a T1 has no Syncro drive, it is still extremely off-road capable. And the waters will flow away sometime.

Nighttime. Stars. Cold. No worries. The wagon train is set up, and everyone is well equipped. With blankets, camping stoves and enthusiasm. Hasta la Shasta! We can do it!

Type
T1
Model year
1965
Built in
Germany
Power (kW/hp)
66/90
Engine (displacement)
1.8 l boxer
Colour
Velvet green
Special features
Camper – always good for a drive through mud

Name
Justin Campbell
Year of birth
1972
Place of birth
Sacramento, USA
Place of residence
Sacramento, USA
Profession
Owner of Kombi Haus VW Auto Shop
Passions
Old Volkswagens, their fans and adventure drives

Refuelling. Checking the engine. Because there is nothing in the wilderness. Just the map, compass, and torch. A tough trip, but the group sticks together. Everyone should arrive

Mud and water splash up metres into the air as Justin Campbell's 1965 Bulli submerges into the quagmire, slides sideways into the depths where it knocks violently against an obstacle with one of its wheels. Violently enough to press the tyre almost up against the rim in one spot. Air escapes, water and muck stream in. With great effort, the Bulli crawls onto dry land. Made it! Immediately, the next VW rumbles up and dives head-first into the morass. And another, and another. The Shasta Snow Trip is underway!

The Shasta Snow Trip, SST for short, is an adventure in the deepest woods of northern California. The mud-dip is designed to set the mood for the coming adventure that takes place at an isolated camp site, which SST veteran Gene Cornelius has christened the Rampaging Spot. Since the first running of the SST 15 years ago, Gene has participated 14 times with his 1960 double cab.

This trip that tests the limits of a T1's durability was thought up by Brian Piercy, a steely Bulli driver and US marine, who incidentally is also a cat lover and one of the best mates you could ever have. Many years ago, Brian was crossing a snow-covered pass in a T1. And the idea took hold in him that other owners of elderly Bullis should also experience this feeling of adventure, independence and achievement. Brian describes the SST this way: "The drive is a bit risky. You should expect the unexpected. An old Bulli adds a certain something to the whole trip. No other make of car can impart so much excitement and enthusiasm."

The field of three participants in the first year has now grown to 50

The three participants in the first year were: Brian, plus Gene, an experienced specialist, and Aaron Boice, who simply wanted to have fun. They were accompanied by three very trusting passengers. Today, up to 50 Bullis make the journey, which can take as long as five days.

The SST participants call themselves newts, i.e. they see themselves as amphibians who feel at home in water. However, to be a fully-fledged newt you have to reach the destination in your own vehicle. Having said that, mutual help along the way is expressly allowed. The newts come from all over the world and have two things in common: their love for adventure and their unbounded, irrational and devoted love for the Bulli. One of them is Justin Campbell. When he was 18, he swapped his 1960 Karmann Ghia for a 1964

panel van. He taught himself how to work on Volkswagens and was soon repairing them for others as well, finding that he could do it "quite well". So he set up his own VW business, the Kombi Haus, in Sacramento, California. "Actually, I just needed a place to keep my VW Bullis and other cars", laughs Justin. "No, seriously, I wanted to offer reliable servicing. To run a business that people can trust."

Trust is something you need on the SST as well, and for that Justin is the perfect choice. For instance, when you have to leave your girlfriend behind on an uphill section in order to reduce weight. Then, Justin can be trusted to take her to Shasta. And if you really get stuck, he has a winch on his Bulli to save the day.

Justin's velvet green 1965 Roadrunner Camper 'Ada' is named after the previous owner. The camper van is a fairly standard one, but it has a few useful extras, such as a kind of roll cage around the outside of the driver's cab and that winch, which is fitted to a bracket at the front and can be mounted on the rear as well. The vehicle is powered by a 1776 cc engine with twin Kadron carburettors. Both Bulli and driver are Shasta veterans.

How having a dog on board saves lives

On his first Shasta trip, Justin took his friend Lou Verrilli along as backup. "It started off with a fright. We didn't like the hotel, so we slept with Louis' dog in the camper. The heater's gas was not turned off and the Bulli started to fill with propane. At first we didn't notice anything, but the dog barked like crazy – and quite possibly saved our lives. 'Welcome to Shasta. You might even die here.' – Lou kept his sense of humour."

The trees reach out with their branches. Pointing to the buses which have advanced very far and want to go further still

Brian Piercy once called the SST an "acceptably risky journey". Fortunately, however, most of the newts have above-average driving skills and to date only one or two Bullis have ended up on their roofs. Nonetheless: your undoing could lie just around the next corner. Many of the roads are winding, narrow and slippery. Vision is often poor as well. Snow, ice, rain, dust, mud and sheer drops, combined with 40 or 50 Bullis, a lack of sleep and mechanical issues seem like the perfect recipe for disaster – but, you see, also for adventure!

Every year there are mechanical breakdowns. Sometimes they are easy to rectify, while on other occasions it takes a bit longer. A couple of years ago, the newly overhauled gearbox of a Bulli driven by Damon

He doesn't shy away from mud driving. Justin Campbell is a relaxed person. Just right for a trip that has more surprises in store than his beard has hairs

Mount Shasta is the destination - a long way and very little time to get there. Yet, it is no hellish trip, but instead a fun tour for people who don't really go for polished paintwork

Dayton, another Shasta veteran, seized up on the first day – just hours before sunrise. The T1 was blocking the mountain pass, and soon the whole convoy of Bullis had come to a stop. Justin unpacked his emergency supplies: a replacement gearbox, an impact wrench and a tool box. Others added more tools, parts and gearbox oil. In no time, the Bulli was jacked up and the engine and gearbox taken out. Before the sun had even risen above the horizon, the VW was ready to go again. It is stories like this that forge a bond between participants and are typical of the Shasta newts.

Sometimes the breakdown even happens before the trip

Richard Kimbrough is a photographer, a newt from way back and rich in SST experience. Sometimes the adventures even start on the journey there. On his 1,200-kilometre journey from Arizona to Stockton, California, he broke down with an engine problem just 60 kilometres from his destination. All that was needed was a couple of calls, then Damon Dayton had found an engine that could be borrowed from a friend, and another mate was on his way to Richard with his trailer. The engine had been unused for decades. But, as we know, old flat engines always work. Just a few hours later, Richard's 1961 Rustybus was up on Damon's ramp. The newts set about their work. One donated a suitable exhaust pipe, another a carburettor. By early evening, a newly fitted, if also slightly less powerful, flat engine was running in the Rustybus. To this day, Richard raves about it: "The engine didn't lose a single drop of oil over the whole distance. But given the long transmission, big, wide tyres and heavy load, the missing horsepower made it hard work for our Bulli." On the first steep uphill approach to Goat Mountain, the adventurers distributed tools, vehicle recovery equipment and heavy luggage to other Bullis and succeeded in getting up and over the top. Then they loaded up the Rustybus again. But the next day the incline was even steeper. Now, the photo equipment was jettisoned as well, and ultimately even girlfriend Marlena Marie switched to Justin's vehicle. The empty Rustybus crept up the mountain, but lost touch with the others. With no tools, vehicle recovery equipment, food, water or camera and incidentally, of course, with no girlfriend, a sense of foreboding crept over Richard. Then smoke appeared from under the dashboard: the electrics had given up the ghost. That was something Richard could repair, but it was getting dark. Then, fortunately, some other newts appeared. He was back in touch with the convoy and drove with the others over Scott Mountain to the destination, Cold Creek Inn in Mount Shasta City.

For Richard it was the Shasta Snow Trip of his life: "After the trip, Marlena and I drove to Justin's place. Thanks to his hospitality, his knowledge and his parts, we were able to put together a new engine from the defective one for the journey home. During the repair work, Justin even let us use his house. And that's typical of the Shasta Snow Trip: it's a trip full of adventure where people make friendships for life." ❤

Unpaved roads, mud and lost screws.
But such moments, such beautiful vistas
make you forget the hardships

Alone as a couple at the end of the world everything begins by the light of the eternal satellite: what could be more wonderful?

GALLERY

ARRIVAL

Of course the journey is the reward – often in the figurative sense as well; but in all honesty: letting the sound of the boxer engine die away in the best place in the world, exchanging the humming in the air for moments of contemplation is actually better. Why continue when one has found a haven of peace. Yet often the destination is not a geographical point at all but an inner one – and one that is actually discovered best of all from the comfort of a T bus. Gentlemen, please stop your engines!

Meadows have never been so lush and green, mountains never so perfectly aligned – that's right, these are the Altai mountains in sunny Siberia

No. No one has ever really experienced this scene – it has simply been copied from a dream

This wonderful book would never have come into being if there hadn't been people who provided their support with love and passion. At the forefront is the team from Volkswagen Commercial Vehicles with Günther Scherelis, Jens Bobsien and Christian Schlüter; and of course all the Bulli lovers around the world who have contributed to this Opus Bulli with their ideas, pictures, contacts, night shifts and friendly words of encouragement. We mustn't forget Mary, who kept topping up the coffee pot, and Bob Dylan for his songs in the background.

Bibliographic information held by the German National Library
The German National Library lists this publication in the German National Bibliography.
Detailed bibliographic data can be found on the internet at http://dnb.dnb.de.

1st Edition
ISBN 978-3-667-10304-8

Idea: Dr Katrin Miele
Concept: Edwin Baaske
Research for stories: Bastian Fuhrmann

Motivator: Dr Nadja Kneissler

Editing: Thorsten Elbrigmann
Text: Edwin Baaske, Catrin Barnsteiner, Renee Brincks, Thorsten Elbrigmann, Bastian Fuhrmann, Arno Hasslinger, Gordon Hempton, Ben Jamin, Dr. Ulrich Kahmann, Richard Kimbrough, Jim Koscs, Jürgen Lewandowski, Kaye Müller, Kazuhiro Nanjo, Bai Ning, Georg Otto, Achim Peitzmeier, Jan van Rossem, Martin Santoro, Christian Schlüter, Shibani Varma, Jean Viljoen, Heiko P. Wacker

Photos: Theodor Barth, Stefan Bau, Markus Bolsinger, Adrian Boot, Jan Bürgermeister, Tony Cenicola/The New York Times/Redux/Laif, Stefen Chow, Thorsten Doerk, Ed Fox, Christian Grund, Markus Heimbach, Markus Hintzen, Erik Hormann, Fotolia.de, Fumihiko Ishii, Ben Jamin, Shouta Kikuchi, Richard Kimbrough, Frank Kayser, Christopher Lamarca, Andreas Lidl, Nina-Maria Oetker, Gordon Ogdon / Walter Productions, Stephan Repke, Camilla Rutherford, Sharleen Rydie, Bengt Stiller, Stephan Szantai, Jean Viljoen, Kari Wilton, Julia Zimmermann, Sven Zimmermann

Art Direction: Jörg Weusthoff and Ralf Reiche
Layout: Weusthoff Noël, Hamburg
Translation: LINGUA. Sprache + mehr…, Barsbüttel
Head of production: Jörn Heese
Lithography: Repromayer, Reutlingen
Printing: Kunst- und Werbedruck, Bad Oeynhausen
Printed in Germany 2015